Transistor Circuit Design Tables

Transistor Circuit Design Tables

DAVID S. TAYLOR, M.A.(OXON), PH.D.
Senior Experimental Officer,
Department of Chemical Engineering and Fuel Technology,
University of Sheffield

LONDON

BUTTERWORTHS

THE BUTTERWORTH GROUP

ENGLAND

Butterworth & Co (Publishers) Ltd
London: 88 Kingsway WC2B 6AB

AUSTRALIA

Butterworth & Co (Australia) Ltd
Sydney: 20 Loftus Street
Melbourne: 343 Little Collins Street
Brisbane: 240 Queen Street

CANADA

Butterworth & Co (Canada) Ltd
Toronto: 14 Curity Avenue 374

NEW ZEALAND

Butterworth & Co (New Zealand) Ltd
Wellington: 49/51 Ballance Street
Auckland: 35 High Street

SOUTH AFRICA

Butterworth & Co (South Africa) (Pty) Ltd
Durban: 33/35 Beach Grove

First published 1971

ISBN 0 408 70146 3

Filmset and printed in England by
Cox & Wyman Ltd
London, Fakenham and Reading

Preface

This book consists of a set of eight tables characterising the properties of components, component combinations, and semiconductor networks containing up to two transistors, such as are likely to be met by transistor circuit designers (whether amateur or professional). The tables were worked out by computation on an I.C.L. 1907 Computer in Extended Mercury Autocode, with input and output on five-track perforated tape. The computer output was programmed to give the final page format, and master copies of the pages were printed on a Creed 54 Teleprinter, prior to being photographed for preparation of the printing plates.

In some cases, the wide range of transistors available makes the preparation of accurate general tables virtually impossible: what has been aimed at is the production of material which will convey a reasonable general guideline for a preliminary design procedure. The accuracy of the tables will depend directly on the tolerances of components used for a circuit (frequently plus or minus 10 per cent) and particularly in the case of the table of common emitter amplifier stages, will also be very sensitive to the characteristics of the transistor involved. In the latter case, discrepancies may be as high as +20 to −30 per cent of the indicated values.

The availability of such tables has already been found to be of great value in paving the way to the establishment of a rapid design procedure for transistor networks, and this publication is the logical outcome of a desire to make the tables available to a wider range of users. Design and computation techniques have been referred continually to random sample measurements made to check a wide range of table predictions where transistors are involved, as a guard against the possible occurrence of large discrepancies between observed and calculated values. Reactions from those who use the tables will be welcomed, both with regard to the choice of contents, and with regard to any circumstances where large errors appear to occur.

Contents

Introduction

This collection of tables was originally compiled for the use of a small Electronics Development Group in the Chemical Engineering and Fuel Technology Department of the University of Sheffield. The idea behind the tables was to produce a kind of 'ready reckoner' for transistor circuit design which would reduce the time spent on the development of d.c. and low frequency (i.e. up to 1 MHz) transistor circuits which had to be designed to fulfil specific functions dictated by the research requirements of members of that Department.

Although the complete theoretical analysis of semiconductor networks needs an accurate knowledge of all the components concerned, in addition to quite elaborate mathematical calculations, the tables were primarily produced as an attempt to provide a guide to the properties of a range of rudimentary circuit groups, with the most significant output properties presented in tabular form.

The underlying idea in the construction of a prototype semiconductor system, therefore, was to produce a specific function by adding together a whole series of basic units, where each unit had properties specified in the tables. Experience shows that this technique saves invaluable time at the 'breadboard' stage even if the guide given by the tables is somewhat approximate in character.

One might, by way of example, consider the problem of the design of the conventional transistor common emitter amplifier stage. All will be familiar with the basic arrangement of four resistors (two to form a potential divider for base bias, one as emitter d.c. feedback, and one as collector load) and one transistor. In the tables, two hundred design examples of this stage are given, giving important properties of the stage such as input resistance, open circuit voltage gain, and examples of overall stage gain when the output of the stage is coupled to the input of another identical one. Other similar situations may spring to the designer's mind. Multivibrators, both astable and monostable, for example, are invaluable as sources of frequency and of pulses of specific length: the tables include nearly eight hundred examples of each, with the relevant output characteristic shown against the components making up the circuit.

Schmitt Trigger circuits, the output of which switches sharply at two specific input voltage levels, are, as many experimenters will appreciate, extremely difficult to design satisfactorily by trial-and-error techniques. In these tables, the computer has been employed on a great deal of otherwise tedious calculation to yield comprehensive sets of Schmitt Trigger circuits formulated in terms of the four resistors which are essentially associated in the circuit, and the two voltage levels at which regenerative switching occurs.

The aim of these tables is to assist the transistor circuit designer by presenting him with a large choice of possible circuit groups with tabulated output properties. In order to prevent the length of tables becoming excessive, restrictions have had to be placed on the range of coverage of each table. However, in many cases, the mathematical behaviour of individual tables is such that if a required characteristic appears to be 'out of range', an accurate estimate can be made by extrapolation either between or beyond those cases actually included, since many sets of figures are related by direct or inverse proportion. In other cases, the tables may be scanned to infer a trend of behaviour which is capable of being at least of qualitative value to the designer.

The interpretation of the figures quoted in the output columns of the tables must be made with reference to the accuracy of the component values used in any particular circumstances. In situations where transistors are not involved, the practical margin of error will be of similar magnitude to the tolerances of the components used. In the case of resistance and capacitance, for example, most components used in practice will usually have 10 per cent tolerance, or possibly 5 per cent at best, and in such cases, only the first one or two significant figures of the output column have any real meaning. The number of significant figures presented in the output columns (which can rise to six in some circumstances) is determined to a considerable extent by the way in which printout instructions can be written into computer programs. The pages of the table are photographed directly from teleprinter versions of the actual computer output tapes, and the available printing space for each output register of the computer has to be defined so that the full variation in the value of that variable may be accommodated throughout the whole program, and this may run through several decades of magnitude, with progressive increase in the number of digits printed in the output columns.

In the case of networks involving transistors, the problem of expected accuracy is more difficult to estimate. It would be possible to carry out exhaustive network analyses involving a transistor of a certain type to produce accurate and predictable results—but this would be of little value to the great majority of people who are much more interested in using their own transistors rather than acquiring certain new types whose properties are tabulated. The tables have been deliberately worked out with this very necessary consideration in mind, and a reasonable compromise has been struck between versatility of the tables and their resultant accuracy. In all cases some responsibility is left with the designer: for example, it is assumed that he will be capable of interpreting all tabulated results to an accuracy compatible with the situation in hand, and that he will be capable of selecting a suitable transistor type (with regard to maximum working voltages, current gain, cutoff frequency, polarity or material), to optimise the output characteristic he is seeking.

As an overall guide, therefore, it is suggested that no figure quoted in these tables may be expected to come within 10 per cent each side of the quoted value in an experimental set-up; certain tables may appear better in practical circumstances, depending on the transistors and components actually used. Even at this degree of predicted accuracy, circuit design can be greatly facilitated if the information is exploited to its fullest extent, since in many cases, a discrepancy occurring with a certain transistor may tend to repeat itself throughout the whole table in a proportional manner, and a correction may be made automatically in subsequent cases.

In many of these tables, the degree of mathematical expertise required to work out the results has been small—the calculations concerned will be familiar to everyone. The value of the computer in such circumstances is primarily for its capability for rapid repetitive calculation, combined with its versatility as a type-setting device, since compilation of such results by hand would be exceptionally tedious, even where the calculation involved is a simple one. Such tables achieve their value by presenting their information content in a readily accessible fashion, and achieve a saving of time that would otherwise be spent making a possible series of calculations.

In other cases, the exhaustiveness of the mathematical treatment given has been equated with the precision with which various parameters might be approached in practice. No calculations pay regard to high-frequency characteristics of transistors and low-frequency characteristics have been correlated with practical measurements on sample germanium and silicon transistors in order to try to arrive at some sort of 'norm' of behaviour in each case. In the cases where transistor current gain is important, such as the performance of common emitter amplifier stages, two separate output tables show how the performance of the stages varies between gains of 40 and 100. In the case of multivibrators and monostable circuits, bias resistors have been chosen to saturate a transistor with gain of 14 or over, and since the time-determining stage of operation of these circuits is in fact an exponential decay occurring with a transistor in a cutoff condition, the output is otherwise independent of transistor gain.

In the case of Schmitt Trigger circuits, the calculations were originally based on a transistor gain spread of 40 to 80, but the circuits should operate satisfactorily with transistors outside this range, since the computer has examined a large number of possible configurations, and has only accepted those which fall very near to the stipulated trip and release voltage requirements.

In all cases involving transistors the tables will work equally well in either the pnp or the npn case, provided the rudimentary precautions of supply (and capacitor) polarities are observed. Drawings have, however, been restricted to the pnp case only for simplicity.

Table 1. Parallel Resistance and Series Capacitance

This table presents values of the 'Parallel Resistance Equation', i.e.

$$\frac{1}{R_p} = \frac{1}{R_1} + \frac{1}{R_2} \qquad \text{or} \qquad R_p = \frac{R_1 R_2}{R_1 + R_2}$$

for a range of combinations based on the preferred scale running from values of 1 000 to 1 000 000. Since the value of two capacitors in series is also expressed by the same type of equation, it may be used for this purpose as well. Provided input and output units are consistent, the actual unit is immaterial and could, for example, be in ohms, picofarads or nanofarads, etc. The table will also hold for any multiplication factor which is common to all three values: thus where the table gives the parallel equivalent value of 1 000 and 6 800 Ω as 872 Ω, we can immediately state that the value of 10 and 68 Ω, in parallel is 8·72 Ω. The table may also be used in reverse: thus if we require a 53 Ω resistor (to 5 per cent), for example, the right-hand column is scanned to locate a parallel combination starting with the digits '5–3'. A suitable possibility is '5 600–100 000' giving 5 303. This would be equivalent to a combination of 56 and 1 000 Ω. Only the 56 Ω component would have to have the appropriate tolerance: the effects of error in the 1 000 component can be estimated by reference again to the table (e.g. a 20 per cent increase, to 1 200, would produce a parallel combination of 535, an increase of only 1 per cent). This procedure also allows the selection of a second resistor to 'trim' the value of one which is initially too high. In such cases the tables show the effect of the parallel addition. If for example, it is required to reduce the nominal value of a 1·2 kΩ resistor by a factor of 2 per cent, the tables show that a reduction of this magnitude would be achieved by the addition of 68 kΩ (giving a hypothetical 1179).

It is also possible to estimate the effective value of three or more resistances in parallel by finding the parallel value of any two components in the group and combining this result in parallel with the next one, till all have been so combined. A certain amount of extrapolation is usually necessary in such cases, because of the presence only of the preferred scale of starting values in the table.

TABLE. 1. VALUES OF PARALLEL RESISTANCE AND SERIES CAPACITANCE.

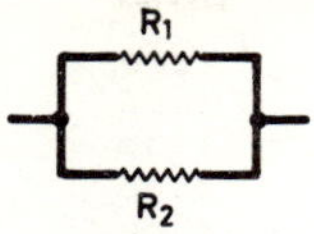

PARALLEL VALUE = R_P

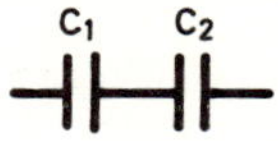

SERIES VALUE = C_S

				PARALLEL RESISTANCE SERIES CAPACITANCE	
R1,C1	R2,C2	RP,CS	R1,C1	R2,C2	RP,CS
1000	1000	500	1200	2200	776
1000	1200	545	1200	2700	831
1000	1500	600	1200	3300	880
1000	1800	643	1200	3900	918
1000	2200	688	1200	4700	956
1000	2700	730	1200	5600	988
1000	3300	767	1200	6800	1020
1000	3900	796	1200	8200	1047
1000	4700	825	1200	10000	1071
1000	5600	848	1200	12000	1091
1000	6800	872	1200	15000	1111
1000	8200	891	1200	18000	1125
1000	10000	909	1200	22000	1138
1000	12000	923	1200	27000	1149
1000	15000	938	1200	33000	1158
1000	18000	947	1200	39000	1164
1000	22000	957	1200	47000	1170
1000	27000	964	1200	56000	1175
1000	33000	971	1200	68000	1179
1000	39000	975	1200	82000	1183
1000	47000	979	1200	100000	1186
1000	56000	982	1200	120000	1188
1000	68000	986	1200	150000	1190
1000	82000	988	1200	180000	1192
1000	100000	990	1200	220000	1193
1000	120000	992	1200	270000	1195
1000	150000	993	1200	330000	1196
1000	180000	994	1200	390000	1196
1000	220000	995	1200	470000	1197
1000	270000	996	1200	560000	1197
1000	330000	997	1200	680000	1198
1000	390000	997	1200	820000	1198
1000	470000	998	1200	1000000	1199
1000	560000	998	1500	1500	750
1000	680000	999	1500	1800	818
1000	820000	999	1500	2200	892
1000	1000000	999	1500	2700	964
1200	1200	600	1500	3300	1031
1200	1500	667	1500	3900	1083
1200	1800	720	1500	4700	1137

PARALLEL RESISTANCE
SERIES CAPACITANCE

R_1,C_1	R_2,C_2	RP,CS	R_1,C_1	R_2,C_2	RP,CS
1500	5600	1183	1800	18000	1636
1500	6800	1229	1800	22000	1664
1500	8200	1268	1800	27000	1688
1500	10000	1304	1800	33000	1707
1500	12000	1333	1800	39000	1721
1500	15000	1364	1800	47000	1734
1500	18000	1385	1800	56000	1744
1500	22000	1404	1800	68000	1754
1500	27000	1421	1800	82000	1761
1500	33000	1435	1800	100000	1768
1500	39000	1444	1800	120000	1773
1500	47000	1454	1800	150000	1779
1500	56000	1461	1800	180000	1782
1500	68000	1468	1800	220000	1785
1500	82000	1473	1800	270000	1788
1500	100000	1478	1800	330000	1790
1500	120000	1481	1800	390000	1792
1500	150000	1485	1800	470000	1793
1500	180000	1488	1800	560000	1794
1500	220000	1490	1800	680000	1795
1500	270000	1492	1800	820000	1796
1500	330000	1493	1800	1000000	1797
1500	390000	1494	2200	2200	1100
1500	470000	1495	2200	2700	1212
1500	560000	1496	2200	3300	1320
1500	680000	1497	2200	3900	1407
1500	820000	1497	2200	4700	1499
1500	1000000	1498	2200	5600	1579
1800	1800	900	2200	6800	1662
1800	2200	990	2200	8200	1735
1800	2700	1080	2200	10000	1803
1800	3300	1165	2200	12000	1859
1800	3900	1232	2200	15000	1919
1800	4700	1302	2200	18000	1960
1800	5600	1362	2200	22000	2000
1800	6800	1423	2200	27000	2034
1800	8200	1476	2200	33000	2063
1800	10000	1525	2200	39000	2083
1800	12000	1565	2200	47000	2102
1800	15000	1607	2200	56000	2117

PARALLEL RESISTANCE
SERIES CAPACITANCE

R1,C1	R2,C2	RP,CS
2200	68000	2131
2200	82000	2143
2200	100000	2153
2200	120000	2160
2200	150000	2168
2200	180000	2173
2200	220000	2178
2200	270000	2182
2200	330000	2185
2200	390000	2188
2200	470000	2190
2200	560000	2191
2200	680000	2193
2200	820000	2194
2200	1000000	2195
2700	2700	1350
2700	3300	1485
2700	3900	1595
2700	4700	1715
2700	5600	1822
2700	6800	1933
2700	8200	2031
2700	10000	2126
2700	12000	2204
2700	15000	2288
2700	18000	2348
2700	22000	2405
2700	27000	2455
2700	33000	2496
2700	39000	2525
2700	47000	2553
2700	56000	2576
2700	68000	2597
2700	82000	2614
2700	100000	2629
2700	120000	2641
2700	150000	2652
2700	180000	2660
2700	220000	2667
2700	270000	2673

R1,C1	R2,C2	RP,CS
2700	330000	2678
2700	390000	2681
2700	470000	2685
2700	560000	2687
2700	680000	2689
2700	820000	2691
2700	1000000	2693
3300	3300	1650
3300	3900	1788
3300	4700	1939
3300	5600	2076
3300	6800	2222
3300	8200	2353
3300	10000	2481
3300	12000	2588
3300	15000	2705
3300	18000	2789
3300	22000	2870
3300	27000	2941
3300	33000	3000
3300	39000	3043
3300	47000	3083
3300	56000	3116
3300	68000	3147
3300	82000	3172
3300	100000	3195
3300	120000	3212
3300	150000	3229
3300	180000	3241
3300	220000	3251
3300	270000	3260
3300	330000	3267
3300	390000	3272
3300	470000	3277
3300	560000	3281
3300	680000	3284
3300	820000	3287
3300	1000000	3289
3900	3900	1950
3900	4700	2131

PARALLEL RESISTANCE
SERIES CAPACITANCE

R1,C1	R2,C2	RP,CS	R1,C1	R2,C2	RP,CS
3900	5600	2299	4700	47000	4273
3900	6800	2479	4700	56000	4336
3900	8200	2643	4700	68000	4396
3900	10000	2806	4700	82000	4445
3900	12000	2943	4700	100000	4489
3900	15000	3095	4700	120000	4523
3900	18000	3205	4700	150000	4557
3900	22000	3313	4700	180000	4580
3900	27000	3408	4700	220000	4602
3900	33000	3488	4700	270000	4620
3900	39000	3545	4700	330000	4634
3900	47000	3601	4700	390000	4644
3900	56000	3646	4700	470000	4653
3900	68000	3688	4700	560000	4661
3900	82000	3723	4700	680000	4668
3900	100000	3754	4700	820000	4673
3900	120000	3777	4700	1000000	4678
3900	150000	3801	5600	5600	2800
3900	180000	3817	5600	6800	3071
3900	220000	3832	5600	8200	3328
3900	270000	3844	5600	10000	3590
3900	330000	3854	5600	12000	3818
3900	390000	3861	5600	15000	4078
3900	470000	3868	5600	18000	4271
3900	560000	3873	5600	22000	4464
3900	680000	3878	5600	27000	4638
3900	820000	3882	5600	33000	4788
3900	1000000	3885	5600	39000	4897
4700	4700	2350	5600	47000	5004
4700	5600	2555	5600	56000	5091
4700	6800	2779	5600	68000	5174
4700	8200	2988	5600	82000	5242
4700	10000	3197	5600	100000	5303
4700	12000	3377	5600	120000	5350
4700	15000	3579	5600	150000	5398
4700	18000	3727	5600	180000	5431
4700	22000	3873	5600	220000	5461
4700	27000	4003	5600	270000	5486
4700	33000	4114	5600	330000	5507
4700	39000	4195	5600	390000	5521

R1,C1	R2,C2	RP,CS	R1,C1	R2,C2	RP,CS
5600	470000	5534	8200	39000	6775
5600	560000	5545	8200	47000	6982
5600	680000	5554	8200	56000	7153
5600	820000	5562	8200	68000	7318
5600	1000000	5569	8200	82000	7455
6800	6800	3400	8200	100000	7579
6800	8200	3717	8200	120000	7676
6800	10000	4048	8200	150000	7775
6800	12000	4340	8200	180000	7843
6800	15000	4679	8200	220000	7905
6800	18000	4935	8200	270000	7958
6800	22000	5194	8200	330000	8001
6800	27000	5432	8200	390000	8031
6800	33000	5638	8200	470000	8059
6800	39000	5790	8200	560000	8082
6800	47000	5941	8200	680000	8102
6800	56000	6064	8200	820000	8119
6800	68000	6182	8200	1000000	8133
6800	82000	6279	10000	10000	5000
6800	100000	6367	10000	12000	5455
6800	120000	6435	10000	15000	6000
6800	150000	6505	10000	18000	6429
6800	180000	6552	10000	22000	6875
6800	220000	6596	10000	27000	7297
6800	270000	6633	10000	33000	7674
6800	330000	6663	10000	39000	7959
6800	390000	6683	10000	47000	8246
6800	470000	6703	10000	56000	8485
6800	560000	6718	10000	68000	8718
6800	680000	6733	10000	82000	8913
6800	820000	6744	10000	100000	9091
6800	1000000	6754	10000	120000	9231
8200	8200	4100	10000	150000	9375
8200	10000	4505	10000	180000	9474
8200	12000	4871	10000	220000	9565
8200	15000	5302	10000	270000	9643
8200	18000	5634	10000	330000	9706
8200	22000	5974	10000	390000	9750
8200	27000	6290	10000	470000	9792
8200	33000	6568	10000	560000	9825

PARALLEL RESISTANCE
SERIES CAPACITANCE

R1,C1	R2,C2	RP,CS	R1,C1	R2,C2	RP,CS
10000	680000	9855	15000	180000	13846
10000	820000	9880	15000	220000	14043
10000	1000000	9901	15000	270000	14211
12000	12000	6000	15000	330000	14348
12000	15000	6667	15000	390000	14444
12000	18000	7200	15000	470000	14536
12000	22000	7765	15000	560000	14609
12000	27000	8308	15000	680000	14676
12000	33000	8800	15000	820000	14731
12000	39000	9176	15000	1000000	14778
12000	47000	9559	18000	18000	9000
12000	56000	9882	18000	22000	9900
12000	68000	10200	18000	27000	10800
12000	82000	10468	18000	33000	11647
12000	100000	10714	18000	39000	12316
12000	120000	10909	18000	47000	13015
12000	150000	11111	18000	56000	13622
12000	180000	11250	18000	68000	14233
12000	220000	11379	18000	82000	14760
12000	270000	11489	18000	100000	15254
12000	330000	11579	18000	120000	15652
12000	390000	11642	18000	150000	16071
12000	470000	11701	18000	180000	16364
12000	560000	11748	18000	220000	16639
12000	680000	11792	18000	270000	16875
12000	820000	11827	18000	330000	17069
12000	1000000	11858	18000	390000	17206
15000	15000	7500	18000	470000	17336
15000	18000	8182	18000	560000	17439
15000	22000	8919	18000	680000	17536
15000	27000	9643	18000	820000	17613
15000	33000	10313	18000	1000000	17682
15000	39000	10833	22000	22000	11000
15000	47000	11371	22000	27000	12122
15000	56000	11831	22000	33000	13200
15000	68000	12289	22000	39000	14066
15000	82000	12680	22000	47000	14986
15000	100000	13043	22000	56000	15795
15000	120000	13333	22000	68000	16622
15000	150000	13636	22000	82000	17346

R_1, C_1	R_2, C_2	RP, CS
22000	100000	18033
22000	120000	18592
22000	150000	19186
22000	180000	19604
22000	220000	20000
22000	270000	20342
22000	330000	20625
22000	390000	20825
22000	470000	21016
22000	560000	21168
22000	680000	21311
22000	820000	21425
22000	1000000	21526
27000	27000	13500
27000	33000	14850
27000	39000	15955
27000	47000	17149
27000	56000	18217
27000	68000	19326
27000	82000	20312
27000	100000	21260
27000	120000	22041
27000	150000	22881
27000	180000	23478
27000	220000	24049
27000	270000	24545
27000	330000	24958
27000	390000	25252
27000	470000	25533
27000	560000	25758
27000	680000	25969
27000	820000	26139
27000	1000000	26290
33000	33000	16500
33000	39000	17875
33000	47000	19388
33000	56000	20764
33000	68000	22218
33000	82000	23530
33000	100000	24812

R_1, C_1	R_2, C_2	RP, CS
33000	120000	25882
33000	150000	27049
33000	180000	27887
33000	220000	28696
33000	270000	29406
33000	330000	30000
33000	390000	30426
33000	470000	30835
33000	560000	31164
33000	680000	31473
33000	820000	31723
33000	1000000	31946
39000	39000	19500
39000	47000	21314
39000	56000	22989
39000	68000	24785
39000	82000	26430
39000	100000	28058
39000	120000	29434
39000	150000	30952
39000	180000	32055
39000	220000	33127
39000	270000	34078
39000	330000	34878
39000	390000	35455
39000	470000	36012
39000	560000	36461
39000	680000	36885
39000	820000	37229
39000	1000000	37536
47000	47000	23500
47000	56000	25553
47000	68000	27791
47000	82000	29876
47000	100000	31973
47000	120000	33772
47000	150000	35787
47000	180000	37269
47000	220000	38727
47000	270000	40032

PARALLEL RESISTANCE
SERIES CAPACITANCE

R_1, C_1	R_2, C_2	RP, CS
47000	330000	41141
47000	390000	41945
47000	470000	42727
47000	560000	43361
47000	680000	43961
47000	820000	44452
47000	1000000	44890
56000	56000	28000
56000	68000	30710
56000	82000	33275
56000	100000	35897
56000	120000	38182
56000	150000	40777
56000	180000	42712
56000	220000	44638
56000	270000	46380
56000	330000	47876
56000	390000	48969
56000	470000	50038
56000	560000	50909
56000	680000	51739
56000	820000	52420
56000	1000000	53030
68000	68000	34000
68000	82000	37173
68000	100000	40476
68000	120000	43404
68000	150000	46789
68000	180000	49355
68000	220000	51944
68000	270000	54320
68000	330000	56382
68000	390000	57904
68000	470000	59405
68000	560000	60637
68000	680000	61818
68000	820000	62793
68000	1000000	63670
82000	82000	41000
82000	100000	45055

R_1, C_1	R_2, C_2	RP, CS
82000	120000	48713
82000	150000	53017
82000	180000	56336
82000	220000	59735
82000	270000	62898
82000	330000	65680
82000	390000	67754
82000	470000	69819
82000	560000	71526
82000	680000	73176
82000	820000	74545
82000	1000000	75786
100000	100000	50000
100000	120000	54545
100000	150000	60000
100000	180000	64286
100000	220000	68750
100000	270000	72973
100000	330000	76744
100000	390000	79592
100000	470000	82456
100000	560000	84848
100000	680000	87179
100000	820000	89130
100000	1000000	90909
120000	120000	60000
120000	150000	66667
120000	180000	72000
120000	220000	77647
120000	270000	83077
120000	330000	88000
120000	390000	91765
120000	470000	95593
120000	560000	98824
120000	680000	102000
120000	820000	104681
120000	1000000	107143
150000	150000	75000
150000	180000	81818
150000	220000	89189

PARALLEL RESISTANCE
SERIES CAPACITANCE

R1,C1	R2,C2	RP,CS	R1,C1	R2,C2	RP,CS
150000	270000	96429	330000	820000	235304
150000	330000	103125	330000	1000000	248120
150000	390000	108333	390000	390000	195000
150000	470000	113710	390000	470000	213140
150000	560000	118310	390000	560000	229895
150000	680000	122892	390000	680000	247850
150000	820000	126804	390000	820000	264298
150000	1000000	130435	390000	1000000	280576
180000	180000	90000	470000	470000	235000
180000	220000	99000	470000	560000	255534
180000	270000	108000	470000	680000	277913
180000	330000	116471	470000	820000	298760
180000	390000	123158	470000	1000000	319728
180000	470000	130154	560000	560000	280000
180000	560000	136216	560000	680000	307097
180000	680000	142326	560000	820000	332754
180000	820000	147600	560000	1000000	358974
180000	1000000	152542	680000	680000	340000
220000	220000	110000	680000	820000	371733
220000	270000	121224	680000	1000000	404762
220000	330000	132000	820000	820000	410000
220000	390000	140656	820000	1000000	450549
220000	470000	149855	1000000	1000000	500000
220000	560000	157949	0	0	0
220000	680000	166222	0	0	0
220000	820000	173462	0	0	0
220000	1000000	180328	0	0	0
270000	270000	135000	0	0	0
270000	330000	148500	0	0	0
270000	390000	159545	0	0	0
270000	470000	171486	0	0	0
270000	560000	182169	0	0	0
270000	680000	193263	0	0	0
270000	820000	203119	0	0	0
270000	1000000	212598	0	0	0
330000	330000	165000	0	0	0
330000	390000	178750	0	0	0
330000	470000	193875	0	0	0
330000	560000	207640	0	0	0
330000	680000	222178	0	0	0

Table 2. Potential Dividers

This table has been compiled with the intention of assisting the choice of two-resistor voltage divider chain primarily for biasing or potential divider applications, such as may occur in differential amplifier stages, stabilised power supplies or in wave shaping circuits where clipping or squaring may be involved. The table takes all preferred combinations of resistance from 1000 to 100 000 Ω between supply rails of 6, 9, and 12 V, and in each case quotes the standing current in the chain in milliamps I(mA), and the potential at the junction (V).

The table behaves in an inversely proportional manner with respect to current: i.e. if both resistance values are increased by a common factor (such as 10, for example), the chain current will decrease by the same factor, although the potential at the junction will remain the same. Both current and voltage values are directly proportional to applied voltage, so multiplication of the quoted supply voltage by any factor will increase both the quoted current and voltage by the same factor. Adjacent tables can also be 'added up'. Thus if it is required to estimate the junction potential for a combination of 1·2 and 8·2 kΩ, between supply rails of 15 V, for example, the answer is the sum of the values in the 6 and 9 V columns, i.e. 0·77+1·15 = 1·92 V. The current in the chain can be estimated in the same way, i.e. 0·64+0·96 = 1·60 mA.

When a potential divider is required to give a precise value (such as is usually required for a stabilised power supply), a preset resistance of low value may be 'sandwiched' between the two resistors of the nearest appropriate chain to provide a degree of variation of potential at the slider. An estimate of the voltage drop across the present may be made by multiplication of the quoted chain current by the resistance of the preset potentiometer.

The choice of chain current for any particular application depends on the magnitude of any further direct currents (from transistor bases, etc.) which are expected to be absorbed by the chain. The potential divider itself will be subject to error in its junction potential as a result of component tolerances and a sensible compromise is to make the ratio of standing to injected current such that the final junction potential will be subject to an error of the same order of magnitude as that caused by component tolerances.

Expressed in practical terms, for 10 per cent component tolerances, the injected current should not exceed approximately 10 per cent of the standing chain current, i.e. the standing chain current should be chosen to be at least ten times as large as any anticipated base currents. For 5 per cent components, the ratio would rise to twenty, and so on.

TABLE. 2. POTENTIAL DIVIDERS.

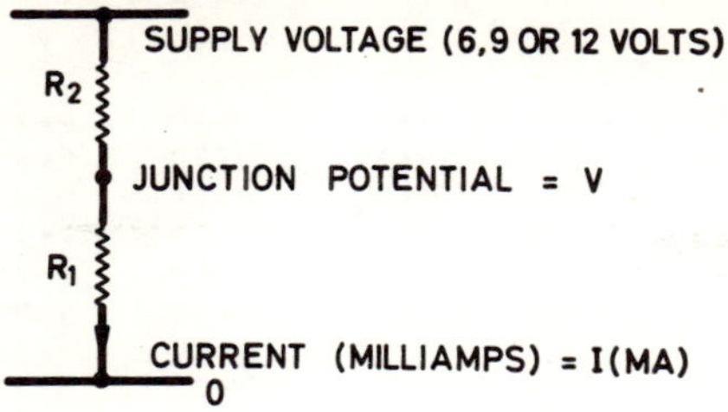

POTENTIAL DIVIDERS

		6 VOLT		9 VOLT		12 VOLT	
R1	R2	I(MA)	V	I(MA)	V	I(MA)	V
1000	1000	3.00	3.00	4.50	4.50	6.00	6.00
1000	1200	2.73	2.73	4.09	4.09	5.45	5.45
1000	1500	2.40	2.40	3.60	3.60	4.80	4.80
1000	1800	2.14	2.14	3.21	3.21	4.29	4.29
1000	2200	1.88	1.88	2.81	2.81	3.75	3.75
1000	2700	1.62	1.62	2.43	2.43	3.24	3.24
1000	3300	1.40	1.40	2.09	2.09	2.79	2.79
1000	3900	1.22	1.22	1.84	1.84	2.45	2.45
1000	4700	1.05	1.05	1.58	1.58	2.11	2.11
1000	5600	0.91	0.91	1.36	1.36	1.82	1.82
1000	6800	0.77	0.77	1.15	1.15	1.54	1.54
1000	8200	0.65	0.65	0.98	0.98	1.30	1.30
1000	10000	0.55	0.55	0.82	0.82	1.09	1.09
1000	12000	0.46	0.46	0.69	0.69	0.92	0.92
1000	15000	0.38	0.38	0.56	0.56	0.75	0.75
1000	18000	0.32	0.32	0.47	0.47	0.63	0.63
1000	22000	0.26	0.26	0.39	0.39	0.52	0.52
1000	27000	0.21	0.21	0.32	0.32	0.43	0.43
1000	33000	0.18	0.18	0.26	0.26	0.35	0.35
1000	39000	0.15	0.15	0.22	0.22	0.30	0.30
1000	47000	0.12	0.12	0.19	0.19	0.25	0.25
1000	56000	0.11	0.11	0.16	0.16	0.21	0.21
1000	68000	0.09	0.09	0.13	0.13	0.17	0.17
1000	82000	0.07	0.07	0.11	0.11	0.14	0.14
1000	100000	0.06	0.06	0.09	0.09	0.12	0.12
1200	1000	2.73	3.27	4.09	4.91	5.45	6.55
1200	1200	2.50	3.00	3.75	4.50	5.00	6.00
1200	1500	2.22	2.67	3.33	4.00	4.44	5.33
1200	1800	2.00	2.40	3.00	3.60	4.00	4.80
1200	2200	1.76	2.12	2.65	3.18	3.53	4.24
1200	2700	1.54	1.85	2.31	2.77	3.08	3.69
1200	3300	1.33	1.60	2.00	2.40	2.67	3.20
1200	3900	1.18	1.41	1.76	2.12	2.35	2.82
1200	4700	1.02	1.22	1.53	1.83	2.03	2.44
1200	5600	0.88	1.06	1.32	1.59	1.76	2.12
1200	6800	0.75	0.90	1.13	1.35	1.50	1.80
1200	8200	0.64	0.77	0.96	1.15	1.28	1.53
1200	10000	0.54	0.64	0.80	0.96	1.07	1.29
1200	12000	0.45	0.55	0.68	0.82	0.91	1.09
1200	15000	0.37	0.44	0.56	0.67	0.74	0.89
1200	18000	0.31	0.38	0.47	0.56	0.62	0.75
1200	22000	0.26	0.31	0.39	0.47	0.52	0.62
1200	27000	0.21	0.26	0.32	0.38	0.43	0.51
1200	33000	0.18	0.21	0.26	0.32	0.35	0.42
1200	39000	0.15	0.18	0.22	0.27	0.30	0.36

POTENTIAL DIVIDERS

R1	R2	6 VOLT I(MA)	6 VOLT V	9 VOLT I(MA)	9 VOLT V	12 VOLT I(MA)	12 VOLT V
1200	47000	0.12	0.15	0.19	0.22	0.25	0.30
1200	56000	0.10	0.13	0.16	0.19	0.21	0.25
1200	68000	0.09	0.10	0.13	0.16	0.17	0.21
1200	82000	0.07	0.09	0.11	0.13	0.14	0.17
1200	100000	0.06	0.07	0.09	0.11	0.12	0.14
1500	1000	2.40	3.60	3.60	5.40	4.80	7.20
1500	1200	2.22	3.33	3.33	5.00	4.44	6.67
1500	1500	2.00	3.00	3.00	4.50	4.00	6.00
1500	1800	1.82	2.73	2.73	4.09	3.64	5.45
1500	2200	1.62	2.43	2.43	3.65	3.24	4.86
1500	2700	1.43	2.14	2.14	3.21	2.86	4.29
1500	3300	1.25	1.88	1.88	2.81	2.50	3.75
1500	3900	1.11	1.67	1.67	2.50	2.22	3.33
1500	4700	0.97	1.45	1.45	2.18	1.94	2.90
1500	5600	0.85	1.27	1.27	1.90	1.69	2.54
1500	6800	0.72	1.08	1.08	1.63	1.45	2.17
1500	8200	0.62	0.93	0.93	1.39	1.24	1.86
1500	10000	0.52	0.78	0.78	1.17	1.04	1.57
1500	12000	0.44	0.67	0.67	1.00	0.89	1.33
1500	15000	0.36	0.55	0.55	0.82	0.73	1.09
1500	18000	0.31	0.46	0.46	0.69	0.62	0.92
1500	22000	0.26	0.38	0.38	0.57	0.51	0.77
1500	27000	0.21	0.32	0.32	0.47	0.42	0.63
1500	33000	0.17	0.26	0.26	0.39	0.35	0.52
1500	39000	0.15	0.22	0.22	0.33	0.30	0.44
1500	47000	0.12	0.19	0.19	0.28	0.25	0.37
1500	56000	0.10	0.16	0.16	0.23	0.21	0.31
1500	68000	0.09	0.13	0.13	0.19	0.17	0.26
1500	82000	0.07	0.11	0.11	0.16	0.14	0.22
1500	100000	0.06	0.09	0.09	0.13	0.12	0.18
1800	1000	2.14	3.86	3.21	5.79	4.29	7.71
1800	1200	2.00	3.60	3.00	5.40	4.00	7.20
1800	1500	1.82	3.27	2.73	4.91	3.64	6.55
1800	1800	1.67	3.00	2.50	4.50	3.33	6.00
1800	2200	1.50	2.70	2.25	4.05	3.00	5.40
1800	2700	1.33	2.40	2.00	3.60	2.67	4.80
1800	3300	1.18	2.12	1.76	3.18	2.35	4.24
1800	3900	1.05	1.89	1.58	2.84	2.11	3.79
1800	4700	0.92	1.66	1.38	2.49	1.85	3.32
1800	5600	0.81	1.46	1.22	2.19	1.62	2.92
1800	6800	0.70	1.26	1.05	1.88	1.40	2.51
1800	8200	0.60	1.08	0.90	1.62	1.20	2.16
1800	10000	0.51	0.92	0.76	1.37	1.02	1.83
1800	12000	0.43	0.78	0.65	1.17	0.87	1.57
1800	15000	0.36	0.64	0.54	0.96	0.71	1.29

		6 VOLT		9 VOLT		12 VOLT	
R1	R2	I(MA)	V	I(MA)	V	I(MA)	V
1800	18000	0.30	0.55	0.45	0.82	0.61	1.09
1800	22000	0.25	0.45	0.38	0.68	0.50	0.91
1800	27000	0.21	0.38	0.31	0.56	0.42	0.75
1800	33000	0.17	0.31	0.26	0.47	0.34	0.62
1800	39000	0.15	0.26	0.22	0.40	0.29	0.53
1800	47000	0.12	0.22	0.18	0.33	0.25	0.44
1800	56000	0.10	0.19	0.16	0.28	0.21	0.37
1800	68000	0.09	0.15	0.13	0.23	0.17	0.31
1800	82000	0.07	0.13	0.11	0.19	0.14	0.26
1800	100000	0.06	0.11	0.09	0.16	0.12	0.21
2200	1000	1.88	4.13	2.81	6.19	3.75	8.25
2200	1200	1.76	3.88	2.65	5.82	3.53	7.76
2200	1500	1.62	3.57	2.43	5.35	3.24	7.14
2200	1800	1.50	3.30	2.25	4.95	3.00	6.60
2200	2200	1.36	3.00	2.05	4.50	2.73	6.00
2200	2700	1.22	2.69	1.84	4.04	2.45	5.39
2200	3300	1.09	2.40	1.64	3.60	2.18	4.80
2200	3900	0.98	2.16	1.48	3.25	1.97	4.33
2200	4700	0.87	1.91	1.30	2.87	1.74	3.83
2200	5600	0.77	1.69	1.15	2.54	1.54	3.38
2200	6800	0.67	1.47	1.00	2.20	1.33	2.93
2200	8200	0.58	1.27	0.87	1.90	1.15	2.54
2200	10000	0.49	1.08	0.74	1.62	0.98	2.16
2200	12000	0.42	0.93	0.63	1.39	0.85	1.86
2200	15000	0.35	0.77	0.52	1.15	0.70	1.53
2200	18000	0.30	0.65	0.45	0.98	0.59	1.31
2200	22000	0.25	0.55	0.37	0.82	0.50	1.09
2200	27000	0.21	0.45	0.31	0.68	0.41	0.90
2200	33000	0.17	0.38	0.26	0.56	0.34	0.75
2200	39000	0.15	0.32	0.22	0.48	0.29	0.64
2200	47000	0.12	0.27	0.18	0.40	0.24	0.54
2200	56000	0.10	0.23	0.15	0.34	0.21	0.45
2200	68000	0.09	0.19	0.13	0.28	0.17	0.38
2200	82000	0.07	0.16	0.11	0.24	0.14	0.31
2200	100000	0.06	0.13	0.09	0.19	0.12	0.26
2700	1000	1.62	4.38	2.43	6.57	3.24	8.76
2700	1200	1.54	4.15	2.31	6.23	3.08	8.31
2700	1500	1.43	3.86	2.14	5.79	2.86	7.71
2700	1800	1.33	3.60	2.00	5.40	2.67	7.20
2700	2200	1.22	3.31	1.84	4.96	2.45	6.61
2700	2700	1.11	3.00	1.67	4.50	2.22	6.00
2700	3300	1.00	2.70	1.50	4.05	2.00	5.40
2700	3900	0.91	2.45	1.36	3.68	1.82	4.91
2700	4700	0.81	2.19	1.22	3.28	1.62	4.38
2700	5600	0.72	1.95	1.08	2.93	1.45	3.90

POTENTIAL DIVIDERS

		6 VOLT		9 VOLT		12 VOLT	
R1	R2	I(MA)	V	I(MA)	V	I(MA)	V
2700	6800	0.63	1.71	0.95	2.56	1.26	3.41
2700	8200	0.55	1.49	0.83	2.23	1.10	2.97
2700	10000	0.47	1.28	0.71	1.91	0.94	2.55
2700	12000	0.41	1.10	0.61	1.65	0.82	2.20
2700	15000	0.34	0.92	0.51	1.37	0.68	1.83
2700	18000	0.29	0.78	0.43	1.17	0.58	1.57
2700	22000	0.24	0.66	0.36	0.98	0.49	1.31
2700	27000	0.20	0.55	0.30	0.82	0.40	1.09
2700	33000	0.17	0.45	0.25	0.68	0.34	0.91
2700	39000	0.14	0.39	0.22	0.58	0.29	0.78
2700	47000	0.12	0.33	0.18	0.49	0.24	0.65
2700	56000	0.10	0.28	0.15	0.41	0.20	0.55
2700	68000	0.08	0.23	0.13	0.34	0.17	0.46
2700	82000	0.07	0.19	0.11	0.29	0.14	0.38
2700	100000	0.06	0.16	0.09	0.24	0.12	0.32
3300	1000	1.40	4.60	2.09	6.91	2.79	9.21
3300	1200	1.33	4.40	2.00	6.60	2.67	8.80
3300	1500	1.25	4.13	1.88	6.19	2.50	8.25
3300	1800	1.18	3.88	1.76	5.82	2.35	7.76
3300	2200	1.09	3.60	1.64	5.40	2.18	7.20
3300	2700	1.00	3.30	1.50	4.95	2.00	6.60
3300	3300	0.91	3.00	1.36	4.50	1.82	6.00
3300	3900	0.83	2.75	1.25	4.13	1.67	5.50
3300	4700	0.75	2.48	1.13	3.71	1.50	4.95
3300	5600	0.67	2.22	1.01	3.34	1.35	4.45
3300	6800	0.59	1.96	0.89	2.94	1.19	3.92
3300	8200	0.52	1.72	0.78	2.58	1.04	3.44
3300	10000	0.45	1.49	0.68	2.23	0.90	2.98
3300	12000	0.39	1.29	0.59	1.94	0.78	2.59
3300	15000	0.33	1.08	0.49	1.62	0.66	2.16
3300	18000	0.28	0.93	0.42	1.39	0.56	1.86
3300	22000	0.24	0.78	0.36	1.17	0.47	1.57
3300	27000	0.20	0.65	0.30	0.98	0.40	1.31
3300	33000	0.17	0.55	0.25	0.82	0.33	1.09
3300	39000	0.14	0.47	0.21	0.70	0.28	0.94
3300	47000	0.12	0.39	0.18	0.59	0.24	0.79
3300	56000	0.10	0.33	0.15	0.50	0.20	0.67
3300	68000	0.08	0.28	0.13	0.42	0.17	0.56
3300	82000	0.07	0.23	0.11	0.35	0.14	0.46
3300	100000	0.06	0.19	0.09	0.29	0.12	0.38
3900	1000	1.22	4.78	1.84	7.16	2.45	9.55
3900	1200	1.18	4.59	1.76	6.88	2.35	9.18
3900	1500	1.11	4.33	1.67	6.50	2.22	8.67
3900	1800	1.05	4.11	1.58	6.16	2.11	8.21
3900	2200	0.98	3.84	1.48	5.75	1.97	7.67

		6 VOLT		9 VOLT		12 VOLT	
R1	R2	I(MA)	V	I(MA)	V	I(MA)	V
3900	2700	0.91	3.55	1.36	5.32	1.82	7.09
3900	3300	0.83	3.25	1.25	4.88	1.67	6.50
3900	3900	0.77	3.00	1.15	4.50	1.54	6.00
3900	4700	0.70	2.72	1.05	4.08	1.40	5.44
3900	5600	0.63	2.46	0.95	3.69	1.26	4.93
3900	6800	0.56	2.19	0.84	3.28	1.12	4.37
3900	8200	0.50	1.93	0.74	2.90	0.99	3.87
3900	10000	0.43	1.68	0.65	2.53	0.86	3.37
3900	12000	0.38	1.47	0.57	2.21	0.75	2.94
3900	15000	0.32	1.24	0.48	1.86	0.63	2.48
3900	18000	0.27	1.07	0.41	1.60	0.55	2.14
3900	22000	0.23	0.90	0.35	1.36	0.46	1.81
3900	27000	0.19	0.76	0.29	1.14	0.39	1.51
3900	33000	0.16	0.63	0.24	0.95	0.33	1.27
3900	39000	0.14	0.55	0.21	0.82	0.28	1.09
3900	47000	0.12	0.46	0.18	0.69	0.24	0.92
3900	56000	0.10	0.39	0.15	0.59	0.20	0.78
3900	68000	0.08	0.33	0.13	0.49	0.17	0.65
3900	82000	0.07	0.27	0.10	0.41	0.14	0.54
3900	100000	0.06	0.23	0.09	0.34	0.12	0.45
4700	1000	1.05	4.95	1.58	7.42	2.11	9.89
4700	1200	1.02	4.78	1.53	7.17	2.03	9.56
4700	1500	0.97	4.55	1.45	6.82	1.94	9.10
4700	1800	0.92	4.34	1.38	6.51	1.85	8.68
4700	2200	0.87	4.09	1.30	6.13	1.74	8.17
4700	2700	0.81	3.81	1.22	5.72	1.62	7.62
4700	3300	0.75	3.53	1.13	5.29	1.50	7.05
4700	3900	0.70	3.28	1.05	4.92	1.40	6.56
4700	4700	0.64	3.00	0.96	4.50	1.28	6.00
4700	5600	0.58	2.74	0.87	4.11	1.17	5.48
4700	6800	0.52	2.45	0.78	3.68	1.04	4.90
4700	8200	0.47	2.19	0.70	3.28	0.93	4.37
4700	10000	0.41	1.92	0.61	2.88	0.82	3.84
4700	12000	0.36	1.69	0.54	2.53	0.72	3.38
4700	15000	0.30	1.43	0.46	2.15	0.61	2.86
4700	18000	0.26	1.24	0.40	1.86	0.53	2.48
4700	22000	0.22	1.06	0.34	1.58	0.45	2.11
4700	27000	0.19	0.89	0.28	1.33	0.38	1.78
4700	33000	0.16	0.75	0.24	1.12	0.32	1.50
4700	39000	0.14	0.65	0.21	0.97	0.27	1.29
4700	47000	0.12	0.55	0.17	0.82	0.23	1.09
4700	56000	0.10	0.46	0.15	0.70	0.20	0.93
4700	68000	0.08	0.39	0.12	0.58	0.17	0.78
4700	82000	0.07	0.33	0.10	0.49	0.14	0.65
4700	100000	0.06	0.27	0.09	0.40	0.11	0.54

POTENTIAL DIVIDERS

R_1	R_2	6 VOLT I(MA)	6 VOLT V	9 VOLT I(MA)	9 VOLT V	12 VOLT I(MA)	12 VOLT V
5600	1000	0.91	5.09	1.36	7.64	1.82	10.18
5600	1200	0.88	4.94	1.32	7.41	1.76	9.88
5600	1500	0.85	4.73	1.27	7.10	1.69	9.46
5600	1800	0.81	4.54	1.22	6.81	1.62	9.08
5600	2200	0.77	4.31	1.15	6.46	1.54	8.62
5600	2700	0.72	4.05	1.08	6.07	1.45	8.10
5600	3300	0.67	3.78	1.01	5.66	1.35	7.55
5600	3900	0.63	3.54	0.95	5.31	1.26	7.07
5600	4700	0.58	3.26	0.87	4.89	1.17	6.52
5600	5600	0.54	3.00	0.80	4.50	1.07	6.00
5600	6800	0.48	2.71	0.73	4.06	0.97	5.42
5600	8200	0.43	2.43	0.65	3.65	0.87	4.87
5600	10000	0.38	2.15	0.58	3.23	0.77	4.31
5600	12000	0.34	1.91	0.51	2.86	0.68	3.82
5600	15000	0.29	1.63	0.44	2.45	0.58	3.26
5600	18000	0.25	1.42	0.38	2.14	0.51	2.85
5600	22000	0.22	1.22	0.33	1.83	0.43	2.43
5600	27000	0.18	1.03	0.28	1.55	0.37	2.06
5600	33000	0.16	0.87	0.23	1.31	0.31	1.74
5600	39000	0.13	0.75	0.20	1.13	0.27	1.51
5600	47000	0.11	0.64	0.17	0.96	0.23	1.28
5600	56000	0.10	0.55	0.15	0.82	0.19	1.09
5600	68000	0.08	0.46	0.12	0.68	0.16	0.91
5600	82000	0.07	0.38	0.10	0.58	0.14	0.77
5600	100000	0.06	0.32	0.09	0.48	0.11	0.64
6800	1000	0.77	5.23	1.15	7.85	1.54	10.46
6800	1200	0.75	5.10	1.13	7.65	1.50	10.20
6800	1500	0.72	4.92	1.08	7.37	1.45	9.83
6800	1800	0.70	4.74	1.05	7.12	1.40	9.49
6800	2200	0.67	4.53	1.00	6.80	1.33	9.07
6800	2700	0.63	4.29	0.95	6.44	1.26	8.59
6800	3300	0.59	4.04	0.89	6.06	1.19	8.08
6800	3900	0.56	3.81	0.84	5.72	1.12	7.63
6800	4700	0.52	3.55	0.78	5.32	1.04	7.10
6800	5600	0.48	3.29	0.73	4.94	0.97	6.58
6800	6800	0.44	3.00	0.66	4.50	0.88	6.00
6800	8200	0.40	2.72	0.60	4.08	0.80	5.44
6800	10000	0.36	2.43	0.54	3.64	0.71	4.86
6800	12000	0.32	2.17	0.48	3.26	0.64	4.34
6800	15000	0.28	1.87	0.41	2.81	0.55	3.74
6800	18000	0.24	1.65	0.36	2.47	0.48	3.29
6800	22000	0.21	1.42	0.31	2.13	0.42	2.83
6800	27000	0.18	1.21	0.27	1.81	0.36	2.41
6800	33000	0.15	1.03	0.23	1.54	0.30	2.05
6800	39000	0.13	0.89	0.20	1.34	0.26	1.78

		6 VOLT		9 VOLT		12 VOLT	
R_1	R_2	I(MA)	V	I(MA)	V	I(MA)	V
6800	47000	0.11	0.76	0.17	1.14	0.22	1.52
6800	56000	0.10	0.65	0.14	0.97	0.19	1.30
6800	68000	0.08	0.55	0.12	0.82	0.16	1.09
6800	82000	0.07	0.46	0.10	0.69	0.14	0.92
6800	100000	0.06	0.38	0.08	0.57	0.11	0.76
8200	1000	0.65	5.35	0.98	8.02	1.30	10.70
8200	1200	0.64	5.23	0.96	7.85	1.28	10.47
8200	1500	0.62	5.07	0.93	7.61	1.24	10.14
8200	1800	0.60	4.92	0.90	7.38	1.20	9.84
8200	2200	0.58	4.73	0.87	7.10	1.15	9.46
8200	2700	0.55	4.51	0.83	6.77	1.10	9.03
8200	3300	0.52	4.28	0.78	6.42	1.04	8.56
8200	3900	0.50	4.07	0.74	6.10	0.99	8.13
8200	4700	0.47	3.81	0.70	5.72	0.93	7.63
8200	5600	0.43	3.57	0.65	5.35	0.87	7.13
8200	6800	0.40	3.28	0.60	4.92	0.80	6.56
8200	8200	0.37	3.00	0.55	4.50	0.73	6.00
8200	10000	0.33	2.70	0.49	4.05	0.66	5.41
8200	12000	0.30	2.44	0.45	3.65	0.59	4.87
8200	15000	0.26	2.12	0.39	3.18	0.52	4.24
8200	18000	0.23	1.88	0.34	2.82	0.46	3.76
8200	22000	0.20	1.63	0.30	2.44	0.40	3.26
8200	27000	0.17	1.40	0.26	2.10	0.34	2.80
8200	33000	0.15	1.19	0.22	1.79	0.29	2.39
8200	39000	0.13	1.04	0.19	1.56	0.25	2.08
8200	47000	0.11	0.89	0.16	1.34	0.22	1.78
8200	56000	0.09	0.77	0.14	1.15	0.19	1.53
8200	68000	0.08	0.65	0.12	0.97	0.16	1.29
8200	82000	0.07	0.55	0.10	0.82	0.13	1.09
8200	100000	0.06	0.45	0.08	0.68	0.11	0.91
10000	1000	0.55	5.45	0.82	8.18	1.09	10.91
10000	1200	0.54	5.36	0.80	8.04	1.07	10.71
10000	1500	0.52	5.22	0.78	7.83	1.04	10.43
10000	1800	0.51	5.08	0.76	7.63	1.02	10.17
10000	2200	0.49	4.92	0.74	7.38	0.98	9.84
10000	2700	0.47	4.72	0.71	7.09	0.94	9.45
10000	3300	0.45	4.51	0.68	6.77	0.90	9.02
10000	3900	0.43	4.32	0.65	6.47	0.86	8.63
10000	4700	0.41	4.08	0.61	6.12	0.82	8.16
10000	5600	0.38	3.85	0.58	5.77	0.77	7.69
10000	6800	0.36	3.57	0.54	5.36	0.71	7.14
10000	8200	0.33	3.30	0.49	4.95	0.66	6.59
10000	10000	0.30	3.00	0.45	4.50	0.60	6.00
10000	12000	0.27	2.73	0.41	4.09	0.55	5.45
10000	15000	0.24	2.40	0.36	3.60	0.48	4.80

POTENTIAL DIVIDERS

		6 VOLT		9 VOLT		12 VOLT	
R_1	R_2	I(MA)	V	I(MA)	V	I(MA)	V
10000	18000	0.21	2.14	0.32	3.21	0.43	4.29
10000	22000	0.19	1.88	0.28	2.81	0.38	3.75
10000	27000	0.16	1.62	0.24	2.43	0.32	3.24
10000	33000	0.14	1.40	0.21	2.09	0.28	2.79
10000	39000	0.12	1.22	0.18	1.84	0.24	2.45
10000	47000	0.11	1.05	0.16	1.58	0.21	2.11
10000	56000	0.09	0.91	0.14	1.36	0.18	1.82
10000	68000	0.08	0.77	0.12	1.15	0.15	1.54
10000	82000	0.07	0.65	0.10	0.98	0.13	1.30
10000	100000	0.05	0.55	0.08	0.82	0.11	1.09
12000	1000	0.46	5.54	0.69	8.31	0.92	11.08
12000	1200	0.45	5.45	0.68	8.18	0.91	10.91
12000	1500	0.44	5.33	0.67	8.00	0.89	10.67
12000	1800	0.43	5.22	0.65	7.83	0.87	10.43
12000	2200	0.42	5.07	0.63	7.61	0.85	10.14
12000	2700	0.41	4.90	0.61	7.35	0.82	9.80
12000	3300	0.39	4.71	0.59	7.06	0.78	9.41
12000	3900	0.38	4.53	0.57	6.79	0.75	9.06
12000	4700	0.36	4.31	0.54	6.47	0.72	8.62
12000	5600	0.34	4.09	0.51	6.14	0.68	8.18
12000	6800	0.32	3.83	0.48	5.74	0.64	7.66
12000	8200	0.30	3.56	0.45	5.35	0.59	7.13
12000	10000	0.27	3.27	0.41	4.91	0.55	6.55
12000	12000	0.25	3.00	0.38	4.50	0.50	6.00
12000	15000	0.22	2.67	0.33	4.00	0.44	5.33
12000	18000	0.20	2.40	0.30	3.60	0.40	4.80
12000	22000	0.18	2.12	0.26	3.18	0.35	4.24
12000	27000	0.15	1.85	0.23	2.77	0.31	3.69
12000	33000	0.13	1.60	0.20	2.40	0.27	3.20
12000	39000	0.12	1.41	0.18	2.12	0.24	2.82
12000	47000	0.10	1.22	0.15	1.83	0.20	2.44
12000	56000	0.09	1.06	0.13	1.59	0.18	2.12
12000	68000	0.08	0.90	0.11	1.35	0.15	1.80
12000	82000	0.06	0.77	0.10	1.15	0.13	1.53
12000	100000	0.05	0.64	0.08	0.96	0.11	1.29
15000	1000	0.38	5.63	0.56	8.44	0.75	11.25
15000	1200	0.37	5.56	0.56	8.33	0.74	11.11
15000	1500	0.36	5.45	0.55	8.18	0.73	10.91
15000	1800	0.36	5.36	0.54	8.04	0.71	10.71
15000	2200	0.35	5.23	0.52	7.85	0.70	10.47
15000	2700	0.34	5.08	0.51	7.63	0.68	10.17
15000	3300	0.33	4.92	0.49	7.38	0.66	9.84
15000	3900	0.32	4.76	0.48	7.14	0.63	9.52
15000	4700	0.30	4.57	0.46	6.85	0.61	9.14
15000	5600	0.29	4.37	0.44	6.55	0.58	8.74

		6 VOLT		9 VOLT		12 VOLT	
R1	R2	I(MA)	V	I(MA)	V	I(MA)	V
15000	6800	0.28	4.13	0.41	6.19	0.55	8.26
15000	8200	0.26	3.88	0.39	5.82	0.52	7.76
15000	10000	0.24	3.60	0.36	5.40	0.48	7.20
15000	12000	0.22	3.33	0.33	5.00	0.44	6.67
15000	15000	0.20	3.00	0.30	4.50	0.40	6.00
15000	18000	0.18	2.73	0.27	4.09	0.36	5.45
15000	22000	0.16	2.43	0.24	3.65	0.32	4.86
15000	27000	0.14	2.14	0.21	3.21	0.29	4.29
15000	33000	0.12	1.88	0.19	2.81	0.25	3.75
15000	39000	0.11	1.67	0.17	2.50	0.22	3.33
15000	47000	0.10	1.45	0.15	2.18	0.19	2.90
15000	56000	0.08	1.27	0.13	1.90	0.17	2.54
15000	68000	0.07	1.08	0.11	1.63	0.14	2.17
15000	82000	0.06	0.93	0.09	1.39	0.12	1.86
15000	100000	0.05	0.78	0.08	1.17	0.10	1.57
18000	1000	0.32	5.68	0.47	8.53	0.63	11.37
18000	1200	0.31	5.63	0.47	8.44	0.62	11.25
18000	1500	0.31	5.54	0.46	8.31	0.62	11.08
18000	1800	0.30	5.45	0.45	8.18	0.61	10.91
18000	2200	0.30	5.35	0.45	8.02	0.59	10.69
18000	2700	0.29	5.22	0.43	7.83	0.58	10.43
18000	3300	0.28	5.07	0.42	7.61	0.56	10.14
18000	3900	0.27	4.93	0.41	7.40	0.55	9.86
18000	4700	0.26	4.76	0.40	7.14	0.53	9.52
18000	5600	0.25	4.58	0.38	6.86	0.51	9.15
18000	6800	0.24	4.35	0.36	6.53	0.48	8.71
18000	8200	0.23	4.12	0.34	6.18	0.46	8.24
18000	10000	0.21	3.86	0.32	5.79	0.43	7.71
18000	12000	0.20	3.60	0.30	5.40	0.40	7.20
18000	15000	0.18	3.27	0.27	4.91	0.36	6.55
18000	18000	0.17	3.00	0.25	4.50	0.33	6.00
18000	22000	0.15	2.70	0.22	4.05	0.30	5.40
18000	27000	0.13	2.40	0.20	3.60	0.27	4.80
18000	33000	0.12	2.12	0.18	3.18	0.24	4.24
18000	39000	0.11	1.89	0.16	2.84	0.21	3.79
18000	47000	0.09	1.66	0.14	2.49	0.18	3.32
18000	56000	0.08	1.46	0.12	2.19	0.16	2.92
18000	68000	0.07	1.26	0.10	1.88	0.14	2.51
18000	82000	0.06	1.08	0.09	1.62	0.12	2.16
18000	100000	0.05	0.92	0.08	1.37	0.10	1.83
22000	1000	0.26	5.74	0.39	8.61	0.52	11.48
22000	1200	0.26	5.69	0.39	8.53	0.52	11.38
22000	1500	0.26	5.62	0.38	8.43	0.51	11.23
22000	1800	0.25	5.55	0.38	8.32	0.50	11.09
22000	2200	0.25	5.45	0.37	8.18	0.50	10.91

POTENTIAL DIVIDERS

		6 VOLT		9 VOLT		12 VOLT	
R1	R2	I(MA)	V	I(MA)	V	I(MA)	V
22000	2700	0.24	5.34	0.36	8.02	0.49	10.69
22000	3300	0.24	5.22	0.36	7.83	0.47	10.43
22000	3900	0.23	5.10	0.35	7.64	0.46	10.19
22000	4700	0.22	4.94	0.34	7.42	0.45	9.89
22000	5600	0.22	4.78	0.33	7.17	0.43	9.57
22000	6800	0.21	4.58	0.31	6.88	0.42	9.17
22000	8200	0.20	4.37	0.30	6.56	0.40	8.74
22000	10000	0.19	4.13	0.28	6.19	0.38	8.25
22000	12000	0.18	3.88	0.26	5.82	0.35	7.76
22000	15000	0.16	3.57	0.24	5.35	0.32	7.14
22000	18000	0.15	3.30	0.22	4.95	0.30	6.60
22000	22000	0.14	3.00	0.20	4.50	0.27	6.00
22000	27000	0.12	2.69	0.18	4.04	0.24	5.39
22000	33000	0.11	2.40	0.16	3.60	0.22	4.80
22000	39000	0.10	2.16	0.15	3.25	0.20	4.33
22000	47000	0.09	1.91	0.13	2.87	0.17	3.83
22000	56000	0.08	1.69	0.12	2.54	0.15	3.38
22000	68000	0.07	1.47	0.10	2.20	0.13	2.93
22000	82000	0.06	1.27	0.09	1.90	0.12	2.54
22000	100000	0.05	1.08	0.07	1.62	0.10	2.16
27000	1000	0.21	5.79	0.32	8.68	0.43	11.57
27000	1200	0.21	5.74	0.32	8.62	0.43	11.49
27000	1500	0.21	5.68	0.32	8.53	0.42	11.37
27000	1800	0.21	5.63	0.31	8.44	0.42	11.25
27000	2200	0.21	5.55	0.31	8.32	0.41	11.10
27000	2700	0.20	5.45	0.30	8.18	0.40	10.91
27000	3300	0.20	5.35	0.30	8.02	0.40	10.69
27000	3900	0.19	5.24	0.29	7.86	0.39	10.49
27000	4700	0.19	5.11	0.28	7.67	0.38	10.22
27000	5600	0.18	4.97	0.28	7.45	0.37	9.94
27000	6800	0.18	4.79	0.27	7.19	0.36	9.59
27000	8200	0.17	4.60	0.26	6.90	0.34	9.20
27000	10000	0.16	4.38	0.24	6.57	0.32	8.76
27000	12000	0.15	4.15	0.23	6.23	0.31	8.31
27000	15000	0.14	3.86	0.21	5.79	0.29	7.71
27000	18000	0.13	3.60	0.20	5.40	0.27	7.20
27000	22000	0.12	3.31	0.18	4.96	0.24	6.61
27000	27000	0.11	3.00	0.17	4.50	0.22	6.00
27000	33000	0.10	2.70	0.15	4.05	0.20	5.40
27000	39000	0.09	2.45	0.14	3.68	0.18	4.91
27000	47000	0.08	2.19	0.12	3.28	0.16	4.38
27000	56000	0.07	1.95	0.11	2.93	0.14	3.90
27000	68000	0.06	1.71	0.09	2.56	0.13	3.41
27000	82000	0.06	1.49	0.08	2.23	0.11	2.97
27000	100000	0.05	1.28	0.07	1.91	0.09	2.55

		6 VOLT		9 VOLT		12 VOLT	
R1	R2	I(MA)	V	I(MA)	V	I(MA)	V
33000	1000	0.18	5.82	0.26	8.74	0.35	11.65
33000	1200	0.18	5.79	0.26	8.68	0.35	11.58
33000	1500	0.17	5.74	0.26	8.61	0.35	11.48
33000	1800	0.17	5.69	0.26	8.53	0.34	11.38
33000	2200	0.17	5.63	0.26	8.44	0.34	11.25
33000	2700	0.17	5.55	0.25	8.32	0.34	11.09
33000	3300	0.17	5.45	0.25	8.18	0.33	10.91
33000	3900	0.16	5.37	0.24	8.05	0.33	10.73
33000	4700	0.16	5.25	0.24	7.88	0.32	10.50
33000	5600	0.16	5.13	0.23	7.69	0.31	10.26
33000	6800	0.15	4.97	0.23	7.46	0.30	9.95
33000	8200	0.15	4.81	0.22	7.21	0.29	9.61
33000	10000	0.14	4.60	0.21	6.91	0.28	9.21
33000	12000	0.13	4.40	0.20	6.60	0.27	8.80
33000	15000	0.12	4.13	0.19	6.19	0.25	8.25
33000	18000	0.12	3.88	0.18	5.82	0.24	7.76
33000	22000	0.11	3.60	0.16	5.40	0.22	7.20
33000	27000	0.10	3.30	0.15	4.95	0.20	6.60
33000	33000	0.09	3.00	0.14	4.50	0.18	6.00
33000	39000	0.08	2.75	0.12	4.13	0.17	5.50
33000	47000	0.08	2.48	0.11	3.71	0.15	4.95
33000	56000	0.07	2.22	0.10	3.34	0.13	4.45
33000	68000	0.06	1.96	0.09	2.94	0.12	3.92
33000	82000	0.05	1.72	0.08	2.58	0.10	3.44
33000	100000	0.05	1.49	0.07	2.23	0.09	2.98
39000	1000	0.15	5.85	0.22	8.78	0.30	11.70
39000	1200	0.15	5.82	0.22	8.73	0.30	11.64
39000	1500	0.15	5.78	0.22	8.67	0.30	11.56
39000	1800	0.15	5.74	0.22	8.60	0.29	11.47
39000	2200	0.15	5.68	0.22	8.52	0.29	11.36
39000	2700	0.14	5.61	0.22	8.42	0.29	11.22
39000	3300	0.14	5.53	0.21	8.30	0.28	11.06
39000	3900	0.14	5.45	0.21	8.18	0.28	10.91
39000	4700	0.14	5.35	0.21	8.03	0.27	10.71
39000	5600	0.13	5.25	0.20	7.87	0.27	10.49
39000	6800	0.13	5.11	0.20	7.66	0.26	10.22
39000	8200	0.13	4.96	0.19	7.44	0.25	9.92
39000	10000	0.12	4.78	0.18	7.16	0.24	9.55
39000	12000	0.12	4.59	0.18	6.88	0.24	9.18
39000	15000	0.11	4.33	0.17	6.50	0.22	8.67
39000	18000	0.11	4.11	0.16	6.16	0.21	8.21
39000	22000	0.10	3.84	0.15	5.75	0.20	7.67
39000	27000	0.09	3.55	0.14	5.32	0.18	7.09
39000	33000	0.08	3.25	0.12	4.88	0.17	6.50
39000	39000	0.08	3.00	0.12	4.50	0.15	6.00

POTENTIAL DIVIDERS

		6 VOLT		9 VOLT		12 VOLT	
R1	R2	I(MA)	V	I(MA)	V	I(MA)	V
39000	47000	0.07	2.72	0.10	4.08	0.14	5.44
39000	56000	0.06	2.46	0.09	3.69	0.13	4.93
39000	68000	0.06	2.19	0.08	3.28	0.11	4.37
39000	82000	0.05	1.93	0.07	2.90	0.10	3.87
39000	100000	0.04	1.68	0.06	2.53	0.09	3.37
47000	1000	0.12	5.88	0.19	8.81	0.25	11.75
47000	1200	0.12	5.85	0.19	8.78	0.25	11.70
47000	1500	0.12	5.81	0.19	8.72	0.25	11.63
47000	1800	0.12	5.78	0.18	8.67	0.25	11.56
47000	2200	0.12	5.73	0.18	8.60	0.24	11.46
47000	2700	0.12	5.67	0.18	8.51	0.24	11.35
47000	3300	0.12	5.61	0.18	8.41	0.24	11.21
47000	3900	0.12	5.54	0.18	8.31	0.24	11.08
47000	4700	0.12	5.45	0.17	8.18	0.23	10.91
47000	5600	0.11	5.36	0.17	8.04	0.23	10.72
47000	6800	0.11	5.24	0.17	7.86	0.22	10.48
47000	8200	0.11	5.11	0.16	7.66	0.22	10.22
47000	10000	0.11	4.95	0.16	7.42	0.21	9.89
47000	12000	0.10	4.78	0.15	7.17	0.20	9.56
47000	15000	0.10	4.55	0.15	6.82	0.19	9.10
47000	18000	0.09	4.34	0.14	6.51	0.18	8.68
47000	22000	0.09	4.09	0.13	6.13	0.17	8.17
47000	27000	0.08	3.81	0.12	5.72	0.16	7.62
47000	33000	0.08	3.53	0.11	5.29	0.15	7.05
47000	39000	0.07	3.28	0.10	4.92	0.14	6.56
47000	47000	0.06	3.00	0.10	4.50	0.13	6.00
47000	56000	0.06	2.74	0.09	4.11	0.12	5.48
47000	68000	0.05	2.45	0.08	3.68	0.10	4.90
47000	82000	0.05	2.19	0.07	3.28	0.09	4.37
47000	100000	0.04	1.92	0.06	2.88	0.08	3.84
56000	1000	0.11	5.89	0.16	8.84	0.21	11.79
56000	1200	0.10	5.87	0.16	8.81	0.21	11.75
56000	1500	0.10	5.84	0.16	8.77	0.21	11.69
56000	1800	0.10	5.81	0.16	8.72	0.21	11.63
56000	2200	0.10	5.77	0.15	8.66	0.21	11.55
56000	2700	0.10	5.72	0.15	8.59	0.20	11.45
56000	3300	0.10	5.67	0.15	8.50	0.20	11.33
56000	3900	0.10	5.61	0.15	8.41	0.20	11.22
56000	4700	0.10	5.54	0.15	8.30	0.20	11.07
56000	5600	0.10	5.45	0.15	8.18	0.19	10.91
56000	6800	0.10	5.35	0.14	8.03	0.19	10.70
56000	8200	0.09	5.23	0.14	7.85	0.19	10.47
56000	10000	0.09	5.09	0.14	7.64	0.18	10.18
56000	12000	0.09	4.94	0.13	7.41	0.18	9.88
56000	15000	0.08	4.73	0.13	7.10	0.17	9.46

		6 VOLT		9 VOLT		12 VOLT	
R1	R2	I(MA)	V	I(MA)	V	I(MA)	V
56000	18000	0.08	4.54	0.12	6.81	0.16	9.08
56000	22000	0.08	4.31	0.12	6.46	0.15	8.62
56000	27000	0.07	4.05	0.11	6.07	0.14	8.10
56000	33000	0.07	3.78	0.10	5.66	0.13	7.55
56000	39000	0.06	3.54	0.09	5.31	0.13	7.07
56000	47000	0.06	3.26	0.09	4.89	0.12	6.52
56000	56000	0.05	3.00	0.08	4.50	0.11	6.00
56000	68000	0.05	2.71	0.07	4.06	0.10	5.42
56000	82000	0.04	2.43	0.07	3.65	0.09	4.87
56000	100000	0.04	2.15	0.06	3.23	0.08	4.31
68000	1000	0.09	5.91	0.13	8.87	0.17	11.83
68000	1200	0.09	5.90	0.13	8.84	0.17	11.79
68000	1500	0.09	5.87	0.13	8.81	0.17	11.74
68000	1800	0.09	5.85	0.13	8.77	0.17	11.69
68000	2200	0.09	5.81	0.13	8.72	0.17	11.62
68000	2700	0.08	5.77	0.13	8.66	0.17	11.54
68000	3300	0.08	5.72	0.13	8.58	0.17	11.44
68000	3900	0.08	5.67	0.13	8.51	0.17	11.35
68000	4700	0.08	5.61	0.12	8.42	0.17	11.22
68000	5600	0.08	5.54	0.12	8.32	0.16	11.09
68000	6800	0.08	5.45	0.12	8.18	0.16	10.91
68000	8200	0.08	5.35	0.12	8.03	0.16	10.71
68000	10000	0.08	5.23	0.12	7.85	0.15	10.46
68000	12000	0.08	5.10	0.11	7.65	0.15	10.20
68000	15000	0.07	4.92	0.11	7.37	0.14	9.83
68000	18000	0.07	4.74	0.10	7.12	0.14	9.49
68000	22000	0.07	4.53	0.10	6.80	0.13	9.07
68000	27000	0.06	4.29	0.09	6.44	0.13	8.59
68000	33000	0.06	4.04	0.09	6.06	0.12	8.08
68000	39000	0.06	3.81	0.08	5.72	0.11	7.63
68000	47000	0.05	3.55	0.08	5.32	0.10	7.10
68000	56000	0.05	3.29	0.07	4.94	0.10	6.58
68000	68000	0.04	3.00	0.07	4.50	0.09	6.00
68000	82000	0.04	2.72	0.06	4.08	0.08	5.44
68000	100000	0.04	2.43	0.05	3.64	0.07	4.86
82000	1000	0.07	5.93	0.11	8.89	0.14	11.86
82000	1200	0.07	5.91	0.11	8.87	0.14	11.83
82000	1500	0.07	5.89	0.11	8.84	0.14	11.78
82000	1800	0.07	5.87	0.11	8.81	0.14	11.74
82000	2200	0.07	5.84	0.11	8.76	0.14	11.69
82000	2700	0.07	5.81	0.11	8.71	0.14	11.62
82000	3300	0.07	5.77	0.11	8.65	0.14	11.54
82000	3900	0.07	5.73	0.10	8.59	0.14	11.46
82000	4700	0.07	5.67	0.10	8.51	0.14	11.35
82000	5600	0.07	5.62	0.10	8.42	0.14	11.23

POTENTIAL DIVIDERS

R1	R2	6 VOLT I(MA)	6 VOLT V	9 VOLT I(MA)	9 VOLT V	12 VOLT I(MA)	12 VOLT V
82000	6800	0.07	5.54	0.10	8.31	0.14	11.08
82000	8200	0.07	5.45	0.10	8.18	0.13	10.91
82000	10000	0.07	5.35	0.10	8.02	0.13	10.70
82000	12000	0.06	5.23	0.10	7.85	0.13	10.47
82000	15000	0.06	5.07	0.09	7.61	0.12	10.14
82000	18000	0.06	4.92	0.09	7.38	0.12	9.84
82000	22000	0.06	4.73	0.09	7.10	0.12	9.46
82000	27000	0.06	4.51	0.08	6.77	0.11	9.03
82000	33000	0.05	4.28	0.08	6.42	0.10	8.56
82000	39000	0.05	4.07	0.07	6.10	0.10	8.13
82000	47000	0.05	3.81	0.07	5.72	0.09	7.63
82000	56000	0.04	3.57	0.07	5.35	0.09	7.13
82000	68000	0.04	3.28	0.06	4.92	0.08	6.56
82000	82000	0.04	3.00	0.05	4.50	0.07	6.00
82000	100000	0.03	2.70	0.05	4.05	0.07	5.41
100000	1000	0.06	5.94	0.09	8.91	0.12	11.88
100000	1200	0.06	5.93	0.09	8.89	0.12	11.86
100000	1500	0.06	5.91	0.09	8.87	0.12	11.82
100000	1800	0.06	5.89	0.09	8.84	0.12	11.79
100000	2200	0.06	5.87	0.09	8.81	0.12	11.74
100000	2700	0.06	5.84	0.09	8.76	0.12	11.68
100000	3300	0.06	5.81	0.09	8.71	0.12	11.62
100000	3900	0.06	5.77	0.09	8.66	0.12	11.55
100000	4700	0.06	5.73	0.09	8.60	0.11	11.46
100000	5600	0.06	5.68	0.09	8.52	0.11	11.36
100000	6800	0.06	5.62	0.08	8.43	0.11	11.24
100000	8200	0.06	5.55	0.08	8.32	0.11	11.09
100000	10000	0.05	5.45	0.08	8.18	0.11	10.91
100000	12000	0.05	5.36	0.08	8.04	0.11	10.71
100000	15000	0.05	5.22	0.08	7.83	0.10	10.43
100000	18000	0.05	5.08	0.08	7.63	0.10	10.17
100000	22000	0.05	4.92	0.07	7.38	0.10	9.84
100000	27000	0.05	4.72	0.07	7.09	0.09	9.45
100000	33000	0.05	4.51	0.07	6.77	0.09	9.02
100000	39000	0.04	4.32	0.06	6.47	0.09	8.63
100000	47000	0.04	4.08	0.06	6.12	0.08	8.16
100000	56000	0.04	3.85	0.06	5.77	0.08	7.69
100000	68000	0.04	3.57	0.05	5.36	0.07	7.14
100000	82000	0.03	3.30	0.05	4.95	0.07	6.59
100000	100000	0.03	3.00	0.05	4.50	0.06	6.00

Table 3. Time Constants

This table gives a range of values of the law of exponential decay which occurs as a characteristic waveform, for example, when a capacitor charges or discharges through a resistor. The tables may be used in several practical instances, one for example being to estimate the time taken to attain a certain voltage drop across a capacitor charging or discharging from a supply rail through a resistor, such as may be met in the development of time delay circuits, or alternatively, in the development of pulse forming (i.e. differentiator) networks, where it is desired to modify the shape of a square wave. The table is divided into two separate sections, each section taking a range of capacitive elements from 1 000 pF to 15 000 pF in combination with fifteen resistance values from 1 to 15 kΩ. Each table quotes the time in microseconds for the initial voltage to fall to a specific fraction of its original value: the two fractions chosen being one-tenth and one-half, respectively.

The mathematics involved is the equation governing the law of exponential decay and may be expressed in the form

$$T = 10^{-6} RC \ln (V_0/V)$$

where T is the time in seconds for the initial voltage across the capacitor C expressed in microfarads (R expressed in ohms) to fall from V_0 to V (where V_0 and V are expressed in the same units). In the present tables, the two ratios chosen for V_0/V are 2 and 10.

If it is necessary to convert for example, a 10 kHz square wave to a pulse train, coupling can be made by a capacitor of unknown value to a specific load resistance (say 5·6 kΩ). It is reasonable to require that the pulses shall have almost decayed well before the voltage steps again (which it will do in 50 μs). The criterion of 'almost decayed' may reasonably be interpreted as having fallen to one-tenth of original voltage, and so the first table would be suitable. If we further assume that this condition should have been reached in 10 μs, the 'one tenth' table is scanned to find the two significant figures of '1–0' corresponding to $R_1 = 5600$. This occurs at $C = 8200$ pF where the corresponding time constant is 106 μs. From the proportional behaviour of the table, we can quote a value one-tenth of this, i.e. 820 pF for a time constant of 10·6 μs. Most practical designers would then prefer 1 000 pF as a somewhat more readily available substitute.

The tables are calculated for the case where C is expressed in picofarads, R_1 in ohms, and the time constant is expressed in microseconds, but since the time constants are directly proportional with regard to both resistance and capacitance, increasing the capacitor unit by a specific factor (e.g. 10^3 or 10^6) will have the same effect on the unit for time constant. Thus in addition to the compatibility of the units picofarad and microsecond, nanofarads are consistent with time constants in milliseconds, and capacitances in microfarads are consistent with time constants in seconds.

If we require, for example, a time delay of 12 s to be obtained by charging a capacitor from a supply rail via a 100 kΩ resistor where the circuit will react when the voltage across the capacitor has risen to approximately half the supply voltage, we look in the 'one-half' table to find where digits '1–0' in the resistance column correspond to digits '1–2' in the time constant column. This occurs at 1 800 pF, where 1 000 Ω gives a constant of 1·2 μs. Converting into compatible units, we can say at once that 1 800 μF and 1 000 Ω will yield a time constant of 1·2 s. Proportionally, 1 800 μF and 100 kΩ will give a 120 s time constant, and consequently, 180 μF and 100 kΩ will have the required constant of 12 s.

TABLE. 3. TIME CONSTANTS.

TC (US)
TIME CONSTANT
IN
MICROSECONDS.

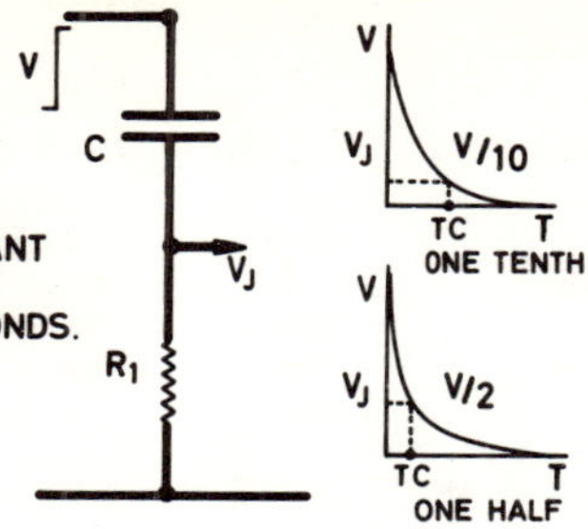

TIME CONSTANTS
ONE TENTH

CAPACITANCE = 1000 PF.

R1	TC(US)	R1	TC(US)	R1	TC(US)
1000	2.3	1200	2.8	1500	3.5
1800	4.1	2200	5.1	2700	6.2
3300	7.6	3900	9.0	4700	10.8
5600	12.9	6800	15.7	8200	18.9
10000	23.0	12000	27.6	15000	34.5

CAPACITANCE = 1200 PF.

R1	TC(US)	R1	TC(US)	R1	TC(US)
1000	2.8	1200	3.3	1500	4.1
1800	5.0	2200	6.1	2700	7.5
3300	9.1	3900	10.8	4700	13.0
5600	15.5	6800	18.8	8200	22.7
10000	27.6	12000	33.2	15000	41.5

CAPACITANCE = 1500 PF.

R1	TC(US)	R1	TC(US)	R1	TC(US)
1000	3.5	1200	4.1	1500	5.2
1800	6.2	2200	7.6	2700	9.3
3300	11.4	3900	13.5	4700	16.2
5600	19.3	6800	23.5	8200	28.3
10000	34.5	12000	41.5	15000	51.8

CAPACITANCE = 1800 PF.

R1	TC(US)	R1	TC(US)	R1	TC(US)
1000	4.1	1200	5.0	1500	6.2
1800	7.5	2200	9.1	2700	11.2
3300	13.7	3900	16.2	4700	19.5
5600	23.2	6800	28.2	8200	34.0
10000	41.5	12000	49.7	15000	62.2

CAPACITANCE = 2200 PF.

R1	TC(US)	R1	TC(US)	R1	TC(US)
1000	5.1	1200	6.1	1500	7.6
1800	9.1	2200	11.1	2700	13.7
3300	16.7	3900	19.8	4700	23.8
5600	28.4	6800	34.5	8200	41.5
10000	50.7	12000	60.8	15000	76.0

TIME CONSTANTS
ONE TENTH

CAPACITANCE = 2700 PF.

R1	TC(US)	R1	TC(US)	R1	TC(US)
1000	6.2	1200	7.5	1500	9.3
1800	11.2	2200	13.7	2700	16.8
3300	20.5	3900	24.3	4700	29.2
5600	34.8	6800	42.3	8200	51.0
10000	62.2	12000	74.6	15000	93.3

CAPACITANCE = 3300 PF.

R1	TC(US)	R1	TC(US)	R1	TC(US)
1000	7.6	1200	9.1	1500	11.4
1800	13.7	2200	16.7	2700	20.5
3300	25.1	3900	29.6	4700	35.7
5600	42.6	6800	51.7	8200	62.3
10000	76.0	12000	91.2	15000	114.0

CAPACITANCE = 3900 PF.

R1	TC(US)	R1	TC(US)	R1	TC(US)
1000	9.0	1200	10.8	1500	13.5
1800	16.2	2200	19.8	2700	24.3
3300	29.6	3900	35.0	4700	42.2
5600	50.3	6800	61.1	8200	73.6
10000	89.8	12000	107.8	15000	134.7

CAPACITANCE = 4700 PF.

R1	TC(US)	R1	TC(US)	R1	TC(US)
1000	10.8	1200	13.0	1500	16.2
1800	19.5	2200	23.8	2700	29.2
3300	35.7	3900	42.2	4700	50.9
5600	60.6	6800	73.6	8200	88.8
10000	108.2	12000	129.9	15000	162.4

CAPACITANCE = 5600 PF.

R1	TC(US)	R1	TC(US)	R1	TC(US)
1000	12.9	1200	15.5	1500	19.3
1800	23.2	2200	28.4	2700	34.8
3300	42.6	3900	50.3	4700	60.6
5600	72.2	6800	87.7	8200	105.8
10000	129.0	12000	154.8	15000	193.5

CAPACITANCE = 6800 PF.

R1	TC(US)	R1	TC(US)	R1	TC(US)
1000	15.7	1200	18.8	1500	23.5
1800	28.2	2200	34.5	2700	42.3
3300	51.7	3900	61.1	4700	73.6
5600	87.7	6800	106.5	8200	128.4
10000	156.6	12000	187.9	15000	234.9

CAPACITANCE = 8200 PF.

R1	TC(US)	R1	TC(US)	R1	TC(US)
1000	18.9	1200	22.7	1500	28.3
1800	34.0	2200	41.5	2700	51.0
3300	62.3	3900	73.6	4700	88.8
5600	105.8	6800	128.4	8200	154.9
10000	188.8	12000	226.6	15000	283.3

CAPACITANCE = 10000 PF.

R1	TC(US)	R1	TC(US)	R1	TC(US)
1000	23.0	1200	27.6	1500	34.5
1800	41.5	2200	50.7	2700	62.2
3300	76.0	3900	89.8	4700	108.2
5600	129.0	6800	156.6	8200	188.8
10000	230.3	12000	276.4	15000	345.4

CAPACITANCE = 12000 PF.

R1	TC(US)	R1	TC(US)	R1	TC(US)
1000	27.6	1200	33.2	1500	41.5
1800	49.7	2200	60.8	2700	74.6
3300	91.2	3900	107.8	4700	129.9
5600	154.8	6800	187.9	8200	226.6
10000	276.4	12000	331.6	15000	414.5

CAPACITANCE = 15000 PF.

R1	TC(US)	R1	TC(US)	R1	TC(US)
1000	34.5	1200	41.5	1500	51.8
1800	62.2	2200	76.0	2700	93.3
3300	114.0	3900	134.7	4700	162.4
5600	193.5	6800	234.9	8200	283.3
10000	345.4	12000	414.5	15000	518.2

TIME CONSTANTS

ONE HALF

CAPACITANCE = 1000 PF.

R1	TC(US)	R1	TC(US)	R1	TC(US)
1000	0.7	1200	0.8	1500	1.0
1800	1.2	2200	1.5	2700	1.9
3300	2.3	3900	2.7	4700	3.3
5600	3.9	6800	4.7	8200	5.7
10000	6.9	12000	8.3	15000	10.4

CAPACITANCE = 1200 PF.

R1	TC(US)	R1	TC(US)	R1	TC(US)
1000	0.8	1200	1.0	1500	1.2
1800	1.5	2200	1.8	2700	2.2
3300	2.7	3900	3.2	4700	3.9
5600	4.7	6800	5.7	8200	6.8
10000	8.3	12000	10.0	15000	12.5

CAPACITANCE = 1500 PF.

R1	TC(US)	R1	TC(US)	R1	TC(US)
1000	1.0	1200	1.2	1500	1.6
1800	1.9	2200	2.3	2700	2.8
3300	3.4	3900	4.1	4700	4.9
5600	5.8	6800	7.1	8200	8.5
10000	10.4	12000	12.5	15000	15.6

CAPACITANCE = 1800 PF.

R1	TC(US)	R1	TC(US)	R1	TC(US)
1000	1.2	1200	1.5	1500	1.9
1800	2.2	2200	2.7	2700	3.4
3300	4.1	3900	4.9	4700	5.9
5600	7.0	6800	8.5	8200	10.2
10000	12.5	12000	15.0	15000	18.7

CAPACITANCE = 2200 PF.

R1	TC(US)	R1	TC(US)	R1	TC(US)
1000	1.5	1200	1.8	1500	2.3
1800	2.7	2200	3.4	2700	4.1
3300	5.0	3900	5.9	4700	7.2
5600	8.5	6800	10.4	8200	12.5
10000	15.2	12000	18.3	15000	22.9

CAPACITANCE = 2700 PF.

R1	TC(US)	R1	TC(US)	R1	TC(US)
1000	1.9	1200	2.2	1500	2.8
1800	3.4	2200	4.1	2700	5.1
3300	6.2	3900	7.3	4700	8.8
5600	10.5	6800	12.7	8200	15.3
10000	18.7	12000	22.5	15000	28.1

CAPACITANCE = 3300 PF.

R1	TC(US)	R1	TC(US)	R1	TC(US)
1000	2.3	1200	2.7	1500	3.4
1800	4.1	2200	5.0	2700	6.2
3300	7.5	3900	8.9	4700	10.7
5600	12.8	6800	15.6	8200	18.8
10000	22.9	12000	27.4	15000	34.3

CAPACITANCE = 3900 PF.

R1	TC(US)	R1	TC(US)	R1	TC(US)
1000	2.7	1200	3.2	1500	4.1
1800	4.9	2200	5.9	2700	7.3
3300	8.9	3900	10.5	4700	12.7
5600	15.1	6800	18.4	8200	22.2
10000	27.0	12000	32.4	15000	40.5

CAPACITANCE = 4700 PF.

R1	TC(US)	R1	TC(US)	R1	TC(US)
1000	3.3	1200	3.9	1500	4.9
1800	5.9	2200	7.2	2700	8.8
3300	10.7	3900	12.7	4700	15.3
5600	18.2	6800	22.2	8200	26.7
10000	32.6	12000	39.1	15000	48.9

CAPACITANCE = 5600 PF.

R1	TC(US)	R1	TC(US)	R1	TC(US)
1000	3.9	1200	4.7	1500	5.8
1800	7.0	2200	8.5	2700	10.5
3300	12.8	3900	15.1	4700	18.2
5600	21.7	6800	26.4	8200	31.8
10000	38.8	12000	46.6	15000	58.2

TIME CONSTANTS
ONE HALF

CAPACITANCE = 6800 PF.

RI	TC(US)	RI	TC(US)	RI	TC(US)
1000	4.7	1200	5.7	1500	7.1
1800	8.5	2200	10.4	2700	12.7
3300	15.6	3900	18.4	4700	22.2
5600	26.4	6800	32.0	8200	38.6
10000	47.1	12000	56.6	15000	70.7

CAPACITANCE = 8200 PF.

RI	TC(US)	RI	TC(US)	RI	TC(US)
1000	5.7	1200	6.8	1500	8.5
1800	10.2	2200	12.5	2700	15.3
3300	18.8	3900	22.2	4700	26.7
5600	31.8	6800	38.6	8200	46.6
10000	56.8	12000	68.2	15000	85.3

CAPACITANCE = 10000 PF.

RI	TC(US)	RI	TC(US)	RI	TC(US)
1000	6.9	1200	8.3	1500	10.4
1800	12.5	2200	15.2	2700	18.7
3300	22.9	3900	27.0	4700	32.6
5600	38.8	6800	47.1	8200	56.8
10000	69.3	12000	83.2	15000	104.0

CAPACITANCE = 12000 PF.

RI	TC(US)	RI	TC(US)	RI	TC(US)
1000	8.3	1200	10.0	1500	12.5
1800	15.0	2200	18.3	2700	22.5
3300	27.4	3900	32.4	4700	39.1
5600	46.6	6800	56.6	8200	68.2
10000	83.2	12000	99.8	15000	124.8

CAPACITANCE = 15000 PF.

RI	TC(US)	RI	TC(US)	RI	TC(US)
1000	10.4	1200	12.5	1500	15.6
1800	18.7	2200	22.9	2700	28.1
3300	34.3	3900	40.5	4700	48.9
5600	58.2	6800	70.7	8200	85.3
10000	104.0	12000	124.8	15000	155.9

Table 4. Capacitor and Inductor Reactances

CAPACITOR REACTANCES

The first of these two tables gives the reactance in ohms of a capacitor over a range of values and at a range of frequencies in accordance with the formula

$$X_c = \frac{10^6}{2\pi FC}$$

where X_c is the reactance in ohms, F is the operating frequency in hertz and C is the capacity in microfarads. This table has been divided into two basic sections, one defining values in the pF–kHz range, and the other in the μF–Hz range. The former table is designed to facilitate selection of capacitors for assembly of resonant networks and for amplifier coupling and decoupling purposes at medium frequency, while the second range is intended for selection of supply rail decoupling capacitors.

The reactance of an amplifier interstage coupling capacitor may be estimated with regard to the input resistance of the following amplifier stage (which can be taken as $R(IN)$ as quoted in Table 5). The capacitor reactance should be small compared with the amplifier input resistance at the lowest anticipated working frequency. A working guideline would be that the reactance should be less than one-tenth of the transistor input resistance. For example, if we require a suitable coupling capacitor for the input of 2 mA common emitter transistor stage operating at 100 kHz, we see that values for $R(IN)$ are in the region of 500 Ω. This would necessitate a reactance of 50 Ω or less at 100 kHz. Such a capacitor, from proportion, would have a reactance of 500 Ω at 10 kHz, and from the table we see that 33 000 pF gives a reactance of 482 Ω, so a 33 000 pF capacitor would be suitable since it would have a reactance of around 48 Ω at 100 kHz.

For selection of an emitter bypass capacitor, a similar argument is applied to the value of the emitter resistor, though the capacitor should have a reactance as low as one fiftieth (i.e. 2 per cent) or less if possible. In the above example, we see that a typical suggested value of emitter resistor is 560 Ω. At 100 kHz, a suitable capacitor would require a reactance about 11 Ω or less, e.g. about 0·15 μF.

When a capacitor has to be chosen for use in a supply rail decoupling arrangement (i.e. a series resistance in the active supply line followed by a capacitor between the supply line and earth), reference may be made to Scroggie's deduction* that the

* M. G. Scroggie, 'Second Thoughts on Radio Theory', p. 193. Iliffe Books Ltd, London (1955).

effectiveness of a capacitor in an R–C decoupling combination may be expressed as an attenuation coefficient, α, such that

$$\alpha = \sqrt{\frac{R^2+X_c^2}{X_c^2}}$$

or more approximately,

$$\alpha \simeq \frac{R}{X_c},$$

the ratio of series resistance to capacitor reactance at the frequency in question. This allows a sensible value for a decoupling capacitor to be estimated in terms of the series resistive term.

If we wish to decouple the supply rail of a d.c. supply to reduce 50 Hz ripple by a factor of 100, for example, and we decide that the maximum tolerable value of series resistance is 2·2 kΩ, we infer that a suitable decoupling capacitor would have a reactance of 22 Ω at 50 Hz. Looking in the table, we observe that at 50 Hz, a capacitor of 15 μF has a reactance of 212 Ω, and hence by extrapolation we may conclude that a 150 μF capacitor would have a reactance of 21·2 Ω, the right order of magnitude for the purpose.

The table sweeps over a range of preferred capacitor values from 1000 to 47000 pF and over a frequency range of 1–15 kHz. In the case of the second range, the values sweep from 1·0 μF (designated by 'MF') to 47 μF for frequencies of 10, 20, 50, 100 and 150 Hz. From the basic reactance equation, it can be seen that there is an inverse proportionality both in terms of frequency and of capacity. Thus if we read from tables that the reactance of an 1800 pF capacitor at 10 kHz is 8842 Ω, we can then say at once a 180 pF capacitor at that frequency has a reactance of 88420 Ω, or that a 1·8 μF capacitor at that frequency has a reactance of 8·842 Ω, since a conversion of units from picofarads to microfarads involves a decrease of 10^6 on reactance where a decrease from 1800 to 1·8 involves an increase of 10^3 on reactance (i.e. shift of the decimal point three places to the left). Alternatively we can say that if the reactance of an 1800 pF capacitor at 10 kHz is 8842 Ω, then at 1 MHz the value is reduced by a factor of 100 to 88·42 Ω, or at 100 Hz the value is increased by a factor of 100 to 884200 Ω.

INDUCTOR REACTANCES

This table is of equivalent purpose to the previous table with the quotation of values of the reactance of an inductor over a range of frequencies, in accordance with the formula

$$X_L = 2\pi FL,$$

where X_L is the reactance in ohms, F is the frequency in hertz, and L is the inductance in henries. The table is divided into two sections, one quoting reactances from 10–470 mH over a frequency range of 1–15 kHz and the other quoting a range of 1–47 H over the range of 10–150 Hz which is primarily for use in low-frequency choke networks. The former table, along with the capacitor reactance table, can be employed for assessing the values of L and C required to produce resonance of a specific frequency by choosing L and C values whose reactances are equal at that frequency. If one is interested in producing a tank circuit resonant at 10 kHz, for example, using an inductor of value 10 mH ($X_L = 628\ \Omega$), we would require a capacitor of the same reactance at 10 kHz. From the table, 27000 pF has $X_c = 589\ \Omega$, and 22000 pF 723 Ω, so a suitable capacitor would have a value around 25000 pF. The reactance of the tank circuit at resonance would not of course, present an apparent resistance of 628 Ω: reactance at resonance depends very much on the 'Q' of the circuit such that R_d, the dynamic resistance of the system

$$R_d = QX,$$

where X is the reactance of the two (equal) reactive members, so that if for example, the 'Q' of the inductor was 10, the dynamic resistance of the assembly at resonance would be 6280 Ω.*

* M. G. Scroggie, 'Second Thoughts on Radio Theory', p. 133. Iliffe Books Ltd, London (1955).

TABLE. 4. CAPACITOR AND INDUCTOR REACTANCES.

F C (μF)

F L (HENRIES)

$$X_C = \frac{10^6}{2\pi FC}$$

$$X_L = 2\pi FL$$

CAPACITOR REACTANCES

FREQUENCY = 1000 HZ

C(PF)	XC	C(PF)	XC	C(PF)	XC
1000	159155	1200	132629	1500	106103
1800	88419	2200	72343	2700	58946
3300	48229	3900	40809	4700	33863
5600	28421	6800	23405	8200	19409
10000	15915	12000	13263	15000	10610
18000	8842	22000	7234	27000	5895
33000	4823	39000	4081	47000	3386

FREQUENCY = 2000 HZ

C(PF)	XC	C(PF)	XC	C(PF)	XC
1000	79577	1200	66315	1500	53052
1800	44210	2200	36172	2700	29473
3300	24114	3900	20404	4700	16931
5600	14210	6800	11703	8200	9705
10000	7958	12000	6631	15000	5305
18000	4421	22000	3617	27000	2947
33000	2411	39000	2040	47000	1693

FREQUENCY = 3000 HZ

C(PF)	XC	C(PF)	XC	C(PF)	XC
1000	53052	1200	44210	1500	35368
1800	29473	2200	24114	2700	19649
3300	16076	3900	13603	4700	11288
5600	9474	6800	7802	8200	6470
10000	5305	12000	4421	15000	3537
18000	2947	22000	2411	27000	1965
33000	1608	39000	1360	47000	1129

FREQUENCY = 4000 HZ

C(PF)	XC	C(PF)	XC	C(PF)	XC
1000	39789	1200	33157	1500	26526
1800	22105	2200	18086	2700	14737
3300	12057	3900	10202	4700	8466
5600	7105	6800	5851	8200	4852
10000	3979	12000	3316	15000	2653
18000	2210	22000	1809	27000	1474
33000	1206	39000	1020	47000	847

FREQUENCY = 5000 HZ

C(PF)	XC	C(PF)	XC	C(PF)	XC
1000	31831	1200	26526	1500	21221
1800	17684	2200	14469	2700	11789
3300	9646	3900	8162	4700	6773
5600	5684	6800	4681	8200	3882
10000	3183	12000	2653	15000	2122
18000	1768	22000	1447	27000	1179
33000	965	39000	816	47000	677

FREQUENCY = 6000 HZ

C(PF)	XC	C(PF)	XC	C(PF)	XC
1000	26526	1200	22105	1500	17684
1800	14737	2200	12057	2700	9824
3300	8038	3900	6801	4700	5644
5600	4737	6800	3901	8200	3235
10000	2653	12000	2210	15000	1768
18000	1474	22000	1206	27000	982
33000	804	39000	680	47000	564

FREQUENCY = 7000 HZ

C(PF)	XC	C(PF)	XC	C(PF)	XC
1000	22736	1200	18947	1500	15158
1800	12631	2200	10335	2700	8421
3300	6890	3900	5830	4700	4838
5600	4060	6800	3344	8200	2773
10000	2274	12000	1895	15000	1516
18000	1263	22000	1033	27000	842
33000	689	39000	583	47000	484

FREQUENCY = 8000 HZ

C(PF)	XC	C(PF)	XC	C(PF)	XC
1000	19894	1200	16579	1500	13263
1800	11052	2200	9043	2700	7368
3300	6029	3900	5101	4700	4233
5600	3553	6800	2926	8200	2426
10000	1989	12000	1658	15000	1326
18000	1105	22000	904	27000	737
33000	603	39000	510	47000	423

FREQUENCY = 9000 HZ

C(PF)	XC	C(PF)	XC	C(PF)	XC
1000	17684	1200	14737	1500	11789
1800	9824	2200	8038	2700	6550
3300	5359	3900	4534	4700	3763
5600	3158	6800	2601	8200	2157
10000	1768	12000	1474	15000	1179
18000	982	22000	804	27000	655
33000	536	39000	453	47000	376

FREQUENCY = 10000 HZ

C(PF)	XC	C(PF)	XC	C(PF)	XC
1000	15915	1200	13263	1500	10610
1800	8842	2200	7234	2700	5895
3300	4823	3900	4081	4700	3386
5600	2842	6800	2341	8200	1941
10000	1592	12000	1326	15000	1061
18000	884	22000	723	27000	589
33000	482	39000	408	47000	339

CAPACITOR REACTANCES

FREQUENCY = 11000 HZ

C(PF)	XC	C(PF)	XC	C(PF)	XC
1000	14469	1200	12057	1500	9646
1800	8038	2200	6577	2700	5359
3300	4384	3900	3710	4700	3078
5600	2584	6800	2128	8200	1764
10000	1447	12000	1206	15000	965
18000	804	22000	658	27000	536
33000	438	39000	371	47000	308

FREQUENCY = 12000 HZ

C(PF)	XC	C(PF)	XC	C(PF)	XC
1000	13263	1200	11052	1500	8842
1800	7368	2200	6029	2700	4912
3300	4019	3900	3401	4700	2822
5600	2368	6800	1950	8200	1617
10000	1326	12000	1105	15000	884
18000	737	22000	603	27000	491
33000	402	39000	340	47000	282

FREQUENCY = 13000 HZ

C(PF)	XC	C(PF)	XC	C(PF)	XC
1000	12243	1200	10202	1500	8162
1800	6801	2200	5565	2700	4534
3300	3710	3900	3139	4700	2605
5600	2186	6800	1800	8200	1493
10000	1224	12000	1020	15000	816
18000	680	22000	556	27000	453
33000	371	39000	314	47000	260

FREQUENCY = 14000 HZ

C(PF)	XC	C(PF)	XC	C(PF)	XC
1000	11368	1200	9474	1500	7579
1800	6316	2200	5167	2700	4210
3300	3445	3900	2915	4700	2419
5600	2030	6800	1672	8200	1386
10000	1137	12000	947	15000	758
18000	632	22000	517	27000	421
33000	344	39000	291	47000	242

FREQUENCY = 15000 HZ

C(PF)	XC	C(PF)	XC	C(PF)	XC
1000	10610	1200	8842	1500	7074
1800	5895	2200	4823	2700	3930
3300	3215	3900	2721	4700	2258
5600	1895	6800	1560	8200	1294
10000	1061	12000	884	15000	707
18000	589	22000	482	27000	393
33000	322	39000	272	47000	226

FREQUENCY = 10 HZ

C(MF)	XC	C(MF)	XC	C(MF)	XC
1.0	15915	1.2	13263	1.5	10610
1.8	8842	2.2	7234	2.7	5895
3.3	4823	3.9	4081	4.7	3386
5.6	2842	6.8	2341	8.2	1941
10.0	1592	12.0	1326	15.0	1061
18.0	884	22.0	723	27.0	589
33.0	482	39.0	408	47.0	339

FREQUENCY = 20 HZ

C(MF)	XC	C(MF)	XC	C(MF)	XC
1.0	7958	1.2	6631	1.5	5305
1.8	4421	2.2	3617	2.7	2947
3.3	2411	3.9	2040	4.7	1693
5.6	1421	6.8	1170	8.2	970
10.0	796	12.0	663	15.0	531
18.0	442	22.0	362	27.0	295
33.0	241	39.0	204	47.0	169

FREQUENCY = 50 HZ

C(MF)	XC	C(MF)	XC	C(MF)	XC
1.0	3183	1.2	2653	1.5	2122
1.8	1768	2.2	1447	2.7	1179
3.3	965	3.9	816	4.7	677
5.6	568	6.8	468	8.2	388
10.0	318	12.0	265	15.0	212
18.0	177	22.0	145	27.0	118
33.0	96	39.0	82	47.0	68

FREQUENCY = 100 HZ

C(MF)	XC	C(MF)	XC	C(MF)	XC
1.0	1592	1.2	1326	1.5	1061
1.8	884	2.2	723	2.7	589
3.3	482	3.9	408	4.7	339
5.6	284	6.8	234	8.2	194
10.0	159	12.0	133	15.0	106
18.0	88	22.0	72	27.0	59
33.0	48	39.0	41	47.0	34

FREQUENCY = 150 HZ

C(MF)	XC	C(MF)	XC	C(MF)	XC
1.0	1061	1.2	884	1.5	707
1.8	589	2.2	482	2.7	393
3.3	322	3.9	272	4.7	226
5.6	189	6.8	156	8.2	129
10.0	106	12.0	88	15.0	71
18.0	59	22.0	48	27.0	39
33.0	32	39.0	27	47.0	23

INDUCTOR REACTANCES

FREQUENCY = 1000 HZ

L(MH)	XL	L(MH)	XL	L(MH)	XL
10	62.8	12	75.4	15	94.2
18	113.1	22	138.2	27	169.6
33	207.3	39	245.0	47	295.3
56	351.9	68	427.3	82	515.2
100	628.3	120	754.0	150	942.5
180	1131.0	220	1382.3	270	1696.5
330	2073.5	390	2450.4	470	2953.1

FREQUENCY = 2000 HZ

L(MH)	XL	L(MH)	XL	L(MH)	XL
10	125.7	12	150.8	15	188.5
18	226.2	22	276.5	27	339.3
33	414.7	39	490.1	47	590.6
56	703.7	68	854.5	82	1030.4
100	1256.6	120	1508.0	150	1885.0
180	2261.9	220	2764.6	270	3392.9
330	4146.9	390	4900.9	470	5906.2

FREQUENCY = 3000 HZ

L(MH)	XL	L(MH)	XL	L(MH)	XL
10	188.5	12	226.2	15	282.7
18	339.3	22	414.7	27	508.9
33	622.0	39	735.1	47	885.9
56	1055.6	68	1281.8	82	1545.7
100	1885.0	120	2261.9	150	2827.4
180	3392.9	220	4146.9	270	5089.4
330	6220.4	390	7351.3	470	8859.3

FREQUENCY = 4000 HZ

L(MH)	XL	L(MH)	XL	L(MH)	XL
10	251.3	12	301.6	15	377.0
18	452.4	22	552.9	27	678.6
33	829.4	39	980.2	47	1181.2
56	1407.4	68	1709.0	82	2060.9
100	2513.3	120	3015.9	150	3769.9
180	4523.9	220	5529.2	270	6785.8
330	8293.8	390	9801.8	470	11812.4

FREQUENCY = 5000 HZ

L(MH)	XL	L(MH)	XL	L(MH)	XL
10	314.2	12	377.0	15	471.2
18	565.5	22	691.2	27	848.2
33	1036.7	39	1225.2	47	1476.5
56	1759.3	68	2136.3	82	2576.1
100	3141.6	120	3769.9	150	4712.4
180	5654.9	220	6911.5	270	8482.3
330	10367.3	390	12252.2	470	14765.5

FREQUENCY = 6000 HZ

L(MH)	XL	L(MH)	XL	L(MH)	XL
10	377.0	12	452.4	15	565.5
18	678.6	22	829.4	27	1017.9
33	1244.1	39	1470.3	47	1771.9
56	2111.2	68	2563.5	82	3091.3
100	3769.9	120	4523.9	150	5654.9
180	6785.8	220	8293.8	270	10178.8
330	12440.7	390	14702.7	470	17718.6

FREQUENCY = 7000 HZ

L(MH)	XL	L(MH)	XL	L(MH)	XL
10	439.8	12	527.8	15	659.7
18	791.7	22	967.6	27	1187.5
33	1451.4	39	1715.3	47	2067.2
56	2463.0	68	2990.8	82	3606.5
100	4398.2	120	5277.9	150	6597.3
180	7916.8	220	9676.1	270	11875.2
330	14514.2	390	17153.1	470	20671.7

FREQUENCY = 8000 HZ

L(MH)	XL	L(MH)	XL	L(MH)	XL
10	502.7	12	603.2	15	754.0
18	904.8	22	1105.8	27	1357.2
33	1658.8	39	1960.4	47	2362.5
56	2814.9	68	3418.1	82	4121.8
100	5026.5	120	6031.9	150	7539.8
180	9047.8	220	11058.4	270	13571.7
330	16587.6	390	19603.5	470	23624.8

FREQUENCY = 9000 HZ

L(MH)	XL	L(MH)	XL	L(MH)	XL
10	565.5	12	678.6	15	848.2
18	1017.9	22	1244.1	27	1526.8
33	1866.1	39	2205.4	47	2657.8
56	3166.7	68	3845.3	82	4637.0
100	5654.9	120	6785.8	150	8482.3
180	10178.8	220	12440.7	270	15268.1
330	18661.1	390	22054.0	470	26577.9

FREQUENCY = 10000 HZ

L(MH)	XL	L(MH)	XL	L(MH)	XL
10	628.3	12	754.0	15	942.5
18	1131.0	22	1382.3	27	1696.5
33	2073.5	39	2450.4	47	2953.1
56	3518.6	68	4272.6	82	5152.2
100	6283.2	120	7539.8	150	9424.8
180	11309.7	220	13823.0	270	16964.6
330	20734.5	390	24504.4	470	29531.0

INDUCTOR REACTANCES

FREQUENCY = 11000 HZ

L(MH)	XL	L(MH)	XL	L(MH)	XL
10	691.2	12	829.4	15	1036.7
18	1244.1	22	1520.5	27	1866.1
33	2280.8	39	2695.5	47	3248.4
56	3870.4	68	4699.8	82	5667.4
100	6911.5	120	8293.8	150	10367.3
180	12440.7	220	15205.3	270	18661.1
330	22808.0	390	26954.9	470	32484.1

FREQUENCY = 12000 HZ

L(MH)	XL	L(MH)	XL	L(MH)	XL
10	754.0	12	904.8	15	1131.0
18	1357.2	22	1658.8	27	2035.8
33	2488.1	39	2940.5	47	3543.7
56	4222.3	68	5127.1	82	6182.7
100	7539.8	120	9047.8	150	11309.7
180	13571.7	220	16587.6	270	20357.5
330	24881.4	390	29405.3	470	35437.2

FREQUENCY = 13000 HZ

L(MH)	XL	L(MH)	XL	L(MH)	XL
10	816.8	12	980.2	15	1225.2
18	1470.3	22	1797.0	27	2205.4
33	2695.5	39	3185.6	47	3839.0
56	4574.2	68	5554.3	82	6697.9
100	8168.1	120	9801.8	150	12252.2
180	14702.7	220	17969.9	270	22054.0
330	26954.9	390	31855.7	470	38390.3

FREQUENCY = 14000 HZ

L(MH)	XL	L(MH)	XL	L(MH)	XL
10	879.6	12	1055.6	15	1319.5
18	1583.4	22	1935.2	27	2375.0
33	2902.8	39	3430.6	47	4134.3
56	4926.0	68	5981.6	82	7213.1
100	8796.5	120	10555.8	150	13194.7
180	15833.6	220	19352.2	270	23750.4
330	29028.3	390	34306.2	470	41343.4

FREQUENCY = 15000 HZ

L(MH)	XL	L(MH)	XL	L(MH)	XL
10	942.5	12	1131.0	15	1413.7
18	1696.5	22	2073.5	27	2544.7
33	3110.2	39	3675.7	47	4429.6
56	5277.9	68	6408.8	82	7728.3
100	9424.8	120	11309.7	150	14137.2
180	16964.6	220	20734.5	270	25446.9
330	31101.8	390	36756.6	470	44296.5

FREQUENCY = 10 HZ

L(H)	XL	L(H)	XL	L(H)	XL
1.0	63	1.2	75	1.5	94
1.8	113	2.2	138	2.7	170
3.3	207	3.9	245	4.7	295
5.6	352	6.8	427	8.2	515
10.0	628	12.0	754	15.0	942
18.0	1131	22.0	1382	27.0	1696
33.0	2073	39.0	2450	47.0	2953

FREQUENCY = 20 HZ

L(H)	XL	L(H)	XL	L(H)	XL
1.0	126	1.2	151	1.5	188
1.8	226	2.2	276	2.7	339
3.3	415	3.9	490	4.7	591
5.6	704	6.8	855	8.2	1030
10.0	1257	12.0	1508	15.0	1885
18.0	2262	22.0	2765	27.0	3393
33.0	4147	39.0	4901	47.0	5906

FREQUENCY = 50 HZ

L(H)	XL	L(H)	XL	L(H)	XL
1.0	314	1.2	377	1.5	471
1.8	565	2.2	691	2.7	848
3.3	1037	3.9	1225	4.7	1477
5.6	1759	6.8	2136	8.2	2576
10.0	3142	12.0	3770	15.0	4712
18.0	5655	22.0	6912	27.0	8482
33.0	10367	39.0	12252	47.0	14765

FREQUENCY = 100 HZ

L(H)	XL	L(H)	XL	L(H)	XL
1.0	628	1.2	754	1.5	942
1.8	1131	2.2	1382	2.7	1696
3.3	2073	3.9	2450	4.7	2953
5.6	3519	6.8	4273	8.2	5152
10.0	6283	12.0	7540	15.0	9425
18.0	11310	22.0	13823	27.0	16965
33.0	20735	39.0	24504	47.0	29531

FREQUENCY = 150 HZ

L(H)	XL	L(H)	XL	L(H)	XL
1.0	942	1.2	1131	1.5	1414
1.8	1696	2.2	2073	2.7	2545
3.3	3110	3.9	3676	4.7	4430
5.6	5278	6.8	6409	8.2	7728
10.0	9425	12.0	11310	15.0	14137
18.0	16965	22.0	20735	27.0	25447
33.0	31102	39.0	36757	47.0	44296

Table 5. Common Emitter Amplifier Stages

The most frequently used component configuration for a transistor amplifier stage is the common emitter arrangement, where the transistor emitter is common to both input and output circuits. Generally, the stage is associated with four resistors, two to establish the quiescent base potential in potential divider arrangement (designated *RBL* and *RBU*), one emitter resistor (*RE*), and a resistor in the collector circuit to act as the load (*RL*). The following table presents a range of such stages both for the purpose of indicating suitable resistor combinations to give a specific range of quiescent collector currents (*IMA*), and for the purpose of making an approximate tabulation of the operating properties of the stage, such as input resistance (*R*(*IN*)), open circuit voltage gain (*VG*), and two further conditions termed 'consecutive gain' (*CG40* and *CG100*), where an indication is given of the practical voltage gain of a stage when its output is loaded by connection to another identical stage. The tables are intended to facilitate the design of an amplifier which may contain one or more common emitter stages, in addition to providing suitable arrangements for a range of constant current sources. The latter find use in differential amplifier circuits where a constant current for the emitter coupled pair may be supplied from the collector of a common emitter stage. The tables present one hundred possible arrangements for both silicon and germanium transistors, five arrangements at five quiescent currents at each of four commonly used supply voltages.

An attempt to predict the input resistance and voltage gain of an amplifier stage is a somewhat speculative process, and simplifying assumptions have to be made before figures such as those presented can be calculated.

The behaviour of a common emitter stage as an amplifier is profoundly influenced by several factors. Junction characteristics vary with type of transistor, ambient temperature, and the performance of the stage as an amplifier depends greatly on the extent to which the emitter resistance is bypassed to alternating currents.

The latter involves the presence of a capacitor (usually an electrolytic) whose reactance will not only vary with operating frequency, but also is subject to wide manufacturing tolerances, in addition to a tendency for the capacity to decrease under the passage of high ripple currents. The calculations therefore make the assumption that the emitter bypass, as well as any interstage coupling capacitors involved, has zero reactance, i.e. infinite capacity. This tends to optimise the gain figures so that in practical systems the measured results are usually lower. Taken as a whole, one might expect variations of up to + 20 per cent and −30 per cent on these tables, which makes them the least accurate of the whole book, but which nevertheless can be adopted as a very useful guide. (For practical systems, a suitable size for an interstage coupling or emitter bypass capacitor may be estimated on the basis of the technique described in the introduction to Table 4.)

In order to calculate the open circuit voltage gain of a common emitter amplifier stage, when the emitter is fully decoupled to alternating currents, it is necessary to establish the change of potential at the collector caused by a specific change of collector current through the load resistance, and compare this with the change of potential at the base caused by the corresponding variation of base current. The collector and base currents bear a fixed ratio to each other for a given transistor, and by experiment, it is possible to produce a pair of equations, one for a silicon emitter–base junction, and the other for germanium which indicate how emitter–base voltage varies with base current:

$$\text{SILICON: } V_{BE} = 0{\cdot}58 + 0{\cdot}026 \ln (i)$$
$$\text{GERMANIUM: } V_{BE} = 0{\cdot}10 + 0{\cdot}023 \ln (i)$$

V_{BE} is the measured emitter–base voltage, and (i) is the base current in microamps. The potential drop is the sum of a fixed threshold of about 0·6 V for silicon and 0·1 V for germanium, with an addition which rises logarithmically with current. (These equations will vary with temperature and type of transistor, and in addition are modified somewhat by the instantaneous voltage drop between transistor emitter and collector. Since this latter varies with the nature of the load resistance which is present when the transistor is acting as an amplifier, the actual equations used for producing the tables were rather more complicated). From a given specific change of collector current, a corresponding change of base current is deduced via the current gain, β, and from the appropriate equation this can be converted to a change of base potential. The change of collector potential is equal to the product of load resistance and collector current change, and the voltage gain of the stage is the ratio of this to the change of base potential. Because of the logarithmic behaviour of the two above equations, the voltage gain of a stage which has a specific collector current is nearly independent of the actual current gain of the transistor, and so the open circuit voltage gain of a stage is quoted as a specific general value (*VG*).

In order to evaluate the input resistance of a common emitter stage, a dynamic calculation is made by creating a small voltage displacement from the quiescent position at the input to the stage, and determining the resulting change in current. (Three components are involved: the two base bias resistors, and the base of the transistor itself, all three being in parallel as far as the measurement is concerned). The previous equations give the voltage-current characteristic for the base, and by differentiation (for the germanium case), we obtain

$$\frac{\mathrm{d}\,V_{BE}}{\mathrm{d}\,i} = \frac{0{\cdot}023}{i}$$

The expression $\mathrm{d}\,V_{BE}/\mathrm{d}\,i$ expresses the potential difference/current ratio for a small displacement and is equivalent to a resistance expressed in volts/microamp. Expressed in ohms:

$$\frac{\mathrm{d}\,V_{BE}}{\mathrm{d}\,I} = \frac{23\,000}{i}\,(\Omega)$$

This infers that the apparent input resistance of a transistor emitter–base junction is an inverse function of the standing base current. If we now consider the input resistance of a transistor common emitter stage with current gain $\beta = 40$ and collector current of 1 mA, we can assume the base current is $1\,000/40 = 25\,\mu$A. Substituting,

$$\frac{\mathrm{d}\,V_{BE}}{\mathrm{d}\,i}\,(\Omega) = \frac{0{\cdot}023 \times 10^6}{25} = 920\,\Omega.$$

This value is reduced somewhat further by the presence of the two bias resistors, both of which are in effect in parallel with the transistor input. The input resistance quoted in the tables (designated *R*(*IN*)) has been calculated on the basis of a transistor of current gain $\beta = 40$ and the input resistance would be very roughly $2\frac{1}{2}$ times as great for the case when $\beta = 100$.

In order to evaluate the overall a.c. voltage gain of an amplifier stage when coupled to a subsequent identical stage, one may assume that the collector load resistor has been replaced by a group of four resistances all in parallel. These are the existing load resistor, the two resistors which form the potential divider bias chain of the next stage, and most important, because it is by far the lowest resistance, the transistor base which is being driven. (The necessary interstage coupling capacitor is assumed to have zero reactance, like the emitter bypass capacitor). Of these four resistances, the parallel value of the first three can be estimated immediately, and the value of the transistor input resistance can be evaluated in the same manner as the previous example. It was shown the input resistance of the transistor depended very much on its base current, examples showing that a 1 mA stage of $\beta = 40$ might have an input resistance around 1 000 Ω, and for $\beta = 100$, the value could approach 2 500 Ω. The apparent voltage gain of the first stage would therefore be greatly dependent on the input resistance, and consequently on the current gain, of the second stage. The first value, (*VG*), represents the voltage gain observable at the unloaded collector of the stage. The second value (*CG*40) indicates the effective voltage gain at the collector of the first stage when the consecutive stage (identical in other respects to the first) has a transistor of current gain $\beta = 40$. The final quoted value (*CG*100) represents the overall stage gain when the consecutive stage has a transistor of current gain $\beta = 100$. The overall voltage gain of an amplifier consisting of two identical common emitter stages with transistor $\beta = 40$ will therefore be the product '*CG*40' for the first stage, and '*VG*' for the subsequent unloaded stage.

TABLE.5. COMMON EMITTER AMPLIFIER STAGES.

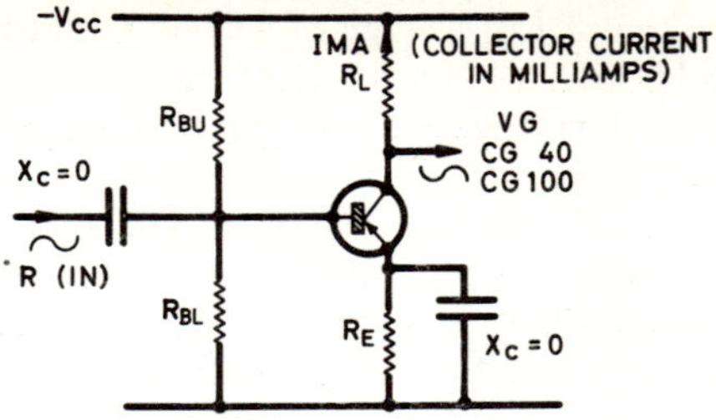

COMMON EMITTER AMPLIFIER STAGES

SILICON 6 VOLT

NOMINAL STAGE CURRENT 0.5 MILLIAMPS
COLLECTOR LOAD RESISTANCE = 5600 OHMS

RBL	RBU	RE	IMA	VG	R(IN)	CG40	CG100
15000	39000	2200	0.45	95	2050	24	36
15000	47000	1800	0.44	95	2170	23	35
10000	33000	1500	0.49	105	1800	23	34
12000	47000	1200	0.46	100	2030	23	34
12000	56000	1000	0.40	85	2330	22	30

NOMINAL STAGE CURRENT 1.0 MILLIAMPS
COLLECTOR LOAD RESISTANCE = 2700 OHMS

RBL	RBU	RE	IMA	VG	R(IN)	CG40	CG100
6800	18000	1000	0.96	95	960	24	36
5600	18000	820	0.91	90	970	23	33
8200	27000	680	1.02	100	1000	25	37
6800	27000	560	0.93	95	1060	23	34
3900	18000	470	0.86	90	960	21	29

NOMINAL STAGE CURRENT 2.0 MILLIAMPS
COLLECTOR LOAD RESISTANCE = 1500 OHMS

RBL	RBU	RE	IMA	VG	R(IN)	CG40	CG100
3300	8200	560	1.83	100	490	24	36
3900	10000	470	2.06	110	470	25	39
3900	12000	390	1.94	105	510	25	38
3300	12000	330	1.79	95	530	24	35
2200	8200	270	2.12	110	420	24	35

NOMINAL STAGE CURRENT 5.0 MILLIAMPS
COLLECTOR LOAD RESISTANCE = 560 OHMS

RBL	RBU	RE	IMA	VG	R(IN)	CG40	CG100
1500	3300	220	5.17	100	180	25	38
1200	3300	180	4.86	95	190	24	36
1500	4700	150	4.77	95	200	24	36
1000	3900	120	4.27	85	200	22	31
820	3300	100	4.81	95	180	22	32

NOMINAL STAGE CURRENT 10.0 MILLIAMPS
COLLECTOR LOAD RESISTANCE = 270 OHMS

RBL	RBU	RE	IMA	VG	R(IN)	CG40	CG100
680	1800	100	9.02	85	100	23	34
470	1500	82	8.58	80	100	22	31
820	2700	68	9.40	85	110	24	35
470	1800	56	9.10	85	100	22	31
680	2700	47	9.67	90	100	23	34

COMMON EMITTER AMPLIFIER STAGES

SILICON 9 VOLT

NOMINAL STAGE CURRENT 0.5 MILLIAMPS
COLLECTOR LOAD RESISTANCE = 8200 OHMS

RBL	RBU	RE	IMA	VG	R(IN)	CG40	CG100
15000	68000	2200	0.43	140	2130	26	42
12000	56000	1800	0.51	160	1810	27	43
10000	56000	1500	0.47	150	1870	25	39
10000	68000	1200	0.42	135	2090	24	36
12000	82000	1000	0.48	155	2020	26	40

NOMINAL STAGE CURRENT 1.0 MILLIAMPS
COLLECTOR LOAD RESISTANCE = 4700 OHMS

RBL	RBU	RE	IMA	VG	R(IN)	CG40	CG100
8200	39000	1000	0.87	155	1100	27	44
6800	33000	820	1.02	175	940	28	45
4700	27000	680	0.96	170	910	26	40
3900	27000	560	0.84	150	970	24	35
4700	33000	470	0.94	165	960	25	39

NOMINAL STAGE CURRENT 2.0 MILLIAMPS
COLLECTOR LOAD RESISTANCE = 2200 OHMS

RBL	RBU	RE	IMA	VG	R(IN)	CG40	CG100
3900	15000	560	2.03	155	470	28	46
3900	18000	470	1.88	150	510	27	44
3300	18000	390	1.77	140	520	26	41
3900	22000	330	1.92	150	520	27	43
3300	22000	270	1.74	140	550	26	39

NOMINAL STAGE CURRENT 5.0 MILLIAMPS
COLLECTOR LOAD RESISTANCE = 820 OHMS

RBL	RBU	RE	IMA	VG	R(IN)	CG40	CG100
1800	6800	220	5.12	140	190	28	46
1200	5600	180	4.77	135	190	26	42
1500	8200	150	4.38	125	220	26	41
1200	6800	120	5.12	140	190	27	42
1000	6800	100	4.36	125	200	25	37

NOMINAL STAGE CURRENT 10.0 MILLIAMPS
COLLECTOR LOAD RESISTANCE = 470 OHMS

RBL	RBU	RE	IMA	VG	R(IN)	CG40	CG100
820	3300	100	10.22	150	100	28	47
680	3300	82	9.56	145	100	27	44
680	3900	68	8.66	135	110	26	41
560	3300	56	9.95	145	90	27	42
390	2700	47	8.70	135	100	24	35

COMMON EMITTER AMPLIFIER STAGES
SILICON 12 VOLT

NOMINAL STAGE CURRENT 0.5 MILLIAMPS
COLLECTOR LOAD RESISTANCE = 12000 OHMS

RBL	RBU	RE	IMA	VG	R(IN)	CG40	CG100
18000	100000	2200	0.51	240	1910	30	52
12000	82000	1800	0.48	230	1880	28	47
15000	120000	1500	0.44	215	2160	28	46
12000	100000	1200	0.51	240	1860	29	47
12000	120000	1000	0.43	210	2180	27	42

NOMINAL STAGE CURRENT 1.0 MILLIAMPS
COLLECTOR LOAD RESISTANCE = 5600 OHMS

RBL	RBU	RE	IMA	VG	R(IN)	CG40	CG100
8200	56000	1000	0.84	185	1100	28	46
5600	39000	820	1.00	210	900	28	46
6800	56000	680	0.89	195	1040	28	45
5600	47000	560	1.04	215	900	28	46
5600	56000	470	0.87	190	1050	27	41

NOMINAL STAGE CURRENT 2.0 MILLIAMPS
COLLECTOR LOAD RESISTANCE = 3300 OHMS

RBL	RBU	RE	IMA	VG	R(IN)	CG40	CG100
4700	27000	560	1.87	215	520	30	52
3300	22000	470	1.82	210	500	29	48
3900	27000	390	2.02	225	490	30	51
2700	22000	330	1.85	215	480	28	45
2700	27000	270	1.49	185	580	26	40

NOMINAL STAGE CURRENT 5.0 MILLIAMPS
COLLECTOR LOAD RESISTANCE = 1200 OHMS

RBL	RBU	RE	IMA	VG	R(IN)	CG40	CG100
1200	6800	220	4.86	190	180	28	48
1800	12000	180	4.52	180	210	29	49
1200	8200	150	5.28	200	180	29	49
1200	10000	120	4.63	185	200	28	45
1200	12000	100	3.75	160	240	26	40

NOMINAL STAGE CURRENT 10.0 MILLIAMPS
COLLECTOR LOAD RESISTANCE = 560 OHMS

RBL	RBU	RE	IMA	VG	R(IN)	CG40	CG100
680	3900	100	10.21	175	90	29	49
680	4700	82	9.30	165	100	28	47
680	5600	68	8.15	150	110	27	43
560	4700	56	9.47	170	100	28	44
560	5600	47	7.62	145	120	26	39

NOMINAL STAGE CURRENT 0.5 MILLIAMPS
COLLECTOR LOAD RESISTANCE = 27000 OHMS

RBL	RBU	RE	IMA	VG	R(IN)	CG40	CG100
12000	150000	2200	0.50	575	1690	32	59
15000	220000	1800	0.47	560	1860	32	59
15000	270000	1500	0.39	505	2140	31	55
12000	220000	1200	0.47	560	1820	31	55
8200	180000	1000	0.40	510	1880	28	47

NOMINAL STAGE CURRENT 1.0 MILLIAMPS
COLLECTOR LOAD RESISTANCE = 12000 OHMS

RBL	RBU	RE	IMA	VG	R(IN)	CG40	CG100
8200	120000	1000	0.84	415	1000	32	58
6800	100000	820	1.00	455	850	32	59
4700	82000	680	0.91	440	850	30	52
5600	100000	560	1.03	465	830	32	56
3900	82000	470	0.89	435	850	29	48

NOMINAL STAGE CURRENT 2.0 MILLIAMPS
COLLECTOR LOAD RESISTANCE = 6800 OHMS

RBL	RBU	RE	IMA	VG	R(IN)	CG40	CG100
3900	47000	560	1.98	425	440	33	62
3300	47000	470	1.83	415	460	32	58
3300	56000	390	1.62	395	510	31	55
2700	47000	330	1.82	415	450	31	54
2200	39000	270	2.14	440	390	31	54

NOMINAL STAGE CURRENT 5.0 MILLIAMPS
COLLECTOR LOAD RESISTANCE = 2700 OHMS

RBL	RBU	RE	IMA	VG	R(IN)	CG40	CG100
1200	15000	220	4.74	360	170	32	58
1500	22000	180	4.39	350	190	32	58
1800	27000	150	4.90	365	190	33	61
1500	27000	120	4.35	350	200	32	56
1200	22000	100	5.01	370	180	32	56

NOMINAL STAGE CURRENT 10.0 MILLIAMPS
COLLECTOR LOAD RESISTANCE = 1200 OHMS

RBL	RBU	RE	IMA	VG	R(IN)	CG40	CG100
820	10000	100	10.38	315	90	33	62
680	10000	82	9.38	305	90	32	58
470	8200	68	8.35	295	90	30	51
560	10000	56	9.35	310	90	31	54
390	8200	47	7.79	290	90	28	45

COMMON EMITTER AMPLIFIER STAGES
GERMANIUM 6 VOLT

NOMINAL STAGE CURRENT 0.5 MILLIAMPS
COLLECTOR LOAD RESISTANCE = 5600 OHMS

RBL	RBU	RE	IMA	VG	R(IN)	CG40	CG100
10000	39000	2200	0.48	85	1930	23	33
8200	39000	1800	0.48	85	1830	22	32
6800	39000	1500	0.48	85	1760	21	30
5600	39000	1200	0.49	85	1660	20	29
5600	47000	1000	0.48	85	1720	20	28

NOMINAL STAGE CURRENT 1.0 MILLIAMPS
COLLECTOR LOAD RESISTANCE = 2700 OHMS

RBL	RBU	RE	IMA	VG	R(IN)	CG40	CG100
6800	27000	1000	1.00	80	1010	24	36
3300	18000	820	0.92	75	900	20	29
3300	22000	680	0.89	75	930	20	28
3900	27000	560	1.01	85	910	22	31
3300	27000	470	1.00	85	890	21	30

NOMINAL STAGE CURRENT 2.0 MILLIAMPS
COLLECTOR LOAD RESISTANCE = 1500 OHMS

RBL	RBU	RE	IMA	VG	R(IN)	CG40	CG100
3300	12000	560	1.92	85	510	24	36
2700	12000	470	1.89	85	500	23	34
2200	12000	390	1.86	85	490	22	32
1800	12000	330	1.78	80	480	21	30
1500	12000	270	1.76	80	460	20	28

NOMINAL SAGE CURRENT 5.0 MILLIAMPS
COLLECTOR LOAD RESISTANCE = 560 OHMS

RBL	RBU	RE	IMA	VG	R(IN)	CG40	CG100
1000	3900	220	4.54	75	200	22	32
1200	5600	180	4.57	75	210	23	33
1000	5600	150	4.54	75	210	22	32
820	5600	120	4.53	75	200	21	30
680	5600	100	4.34	75	200	20	28

NOMINAL STAGE CURRENT 10.0 MILLIAMPS
COLLECTOR LOAD RESISTANCE = 270 OHMS

RBL	RBU	RE	IMA	VG	R(IN)	CG40	CG100
680	2700	100	9.48	75	100	23	35
560	2700	82	9.50	75	100	23	33
470	2700	68	9.48	75	100	22	32
390	2700	56	9.27	75	100	21	30
270	2200	47	9.10	75	90	19	26

NOMINAL STAGE CURRENT 0.5 MILLIAMPS
COLLECTOR LOAD RESISTANCE = 8200 OHMS

RBL	RBU	RE	IMA	VG	R(IN)	CG40	CG100
10000	68000	2200	0.44	115	2050	24	38
10000	82000	1800	0.45	115	2080	24	37
8200	82000	1500	0.43	110	2030	23	35
4700	56000	1200	0.45	115	1650	21	29
5600	68000	1000	0.52	130	1620	23	34

NOMINAL STAGE CURRENT 1.0 MILLIAMPS
COLLECTOR LOAD RESISTANCE = 4700 OHMS

RBL	RBU	RE	IMA	VG	R(IN)	CG40	CG100
3900	27000	1000	0.95	130	920	24	38
3300	27000	820	0.97	130	870	24	36
3900	39000	680	0.93	130	960	24	37
3300	39000	560	0.93	130	920	23	35
3300	47000	470	0.87	125	980	23	33

NOMINAL STAGE CURRENT 2.0 MILLIAMPS
COLLECTOR LOAD RESISTANCE = 2200 OHMS

RBL	RBU	RE	IMA	VG	R(IN)	CG40	CG100
3900	22000	560	2.00	120	520	27	44
2200	15000	470	2.00	125	460	25	40
1800	15000	390	1.95	120	440	24	37
2200	22000	330	1.85	120	500	25	37
1500	18000	270	1.84	120	450	22	33

NOMINAL STAGE CURRENT 5.0 MILLIAMPS
COLLECTOR LOAD RESISTANCE = 820 OHMS

RBL	RBU	RE	IMA	VG	R(IN)	CG40	CG100
1200	6800	220	5.03	110	190	26	42
1200	8200	180	5.02	110	200	26	41
820	6800	150	4.93	110	180	25	37
820	8200	120	4.91	110	190	24	37
680	8200	100	4.69	110	190	23	34

NOMINAL STAGE CURRENT 10.0 MILLIAMPS
COLLECTOR LOAD RESISTANCE = 470 OHMS

RBL	RBU	RE	IMA	VG	R(IN)	CG40	CG100
390	2700	100	8.99	110	90	24	37
330	2700	82	9.11	115	90	23	35
390	3300	68	10.38	125	90	25	39
390	3900	56	10.23	125	90	25	38
270	3300	47	9.56	120	80	22	32

COMMON EMITTER AMPLIFIER STAGES
GERMANIUM 12 VOLT

NOMINAL STAGE CURRENT 0.5 MILLIAMPS
COLLECTOR LOAD RESISTANCE = 12000 OHMS

RBL	RBU	RE	IMA	VG	R(IN)	CG40	CG100
12000	100000	2200	0.50	170	1950	28	47
12000	120000	1800	0.50	175	1980	28	46
8200	100000	1500	0.49	170	1850	26	42
6800	100000	1200	0.49	175	1760	25	39
5600	100000	1000	0.47	170	1720	24	35

NOMINAL STAGE CURRENT 1.0 MILLIAMPS
COLLECTOR LOAD RESISTANCE = 5600 OHMS

RBL	RBU	RE	IMA	VG	R(IN)	CG40	CG100
5600	56000	1000	0.89	145	1050	27	43
3300	39000	820	0.92	150	890	24	37
3300	47000	680	0.89	145	920	24	36
3300	56000	560	0.87	145	950	24	35
2200	39000	470	0.99	160	760	22	32

NOMINAL STAGE CURRENT 2.0 MILLIAMPS
COLLECTOR LOAD RESISTANCE = 3300 OHMS

RBL	RBU	RE	IMA	VG	R(IN)	CG40	CG100
3300	27000	560	1.93	165	510	29	48
2700	27000	470	1.86	165	510	27	45
2700	33000	390	1.78	160	530	27	43
1500	22000	330	1.74	160	460	24	36
2200	33000	270	1.99	170	480	27	42

NOMINAL STAGE CURRENT 5.0 MILLIAMPS
COLLECTOR LOAD RESISTANCE = 1200 OHMS

RBL	RBU	RE	IMA	VG	R(IN)	CG40	CG100
1200	10000	220	4.74	145	200	28	46
1200	12000	180	4.71	145	210	28	45
1000	12000	150	4.60	145	200	27	42
820	12000	120	4.52	145	200	25	39
470	8200	100	4.41	145	170	22	32

NOMINAL STAGE CURRENT 10.0 MILLIAMPS
COLLECTOR LOAD RESISTANCE = 560 OHMS

RBL	RBU	RE	IMA	VG	R(IN)	CG40	CG100
560	4700	100	10.18	140	90	28	46
560	5600	82	10.13	140	100	28	45
390	4700	68	9.93	140	90	26	41
270	3900	56	9.73	140	80	24	35
330	5600	47	9.20	135	90	24	36

COMMON EMITTER AMPLIFIER STAGES
GERMANIUM 24 VOLT

NOMINAL STAGE CURRENT 0.5 MILLIAMPS
COLLECTOR LOAD RESISTANCE = 27000 OHMS

RBL	RBU	RE	IMA	VG	R(IN)	CG40	CG100
12000	220000	2200	0.48	320	1930	31	55
10000	220000	1800	0.48	325	1870	30	52
6800	180000	1500	0.47	320	1730	28	46
5600	180000	1200	0.47	325	1650	26	42
4700	180000	1000	0.45	320	1600	25	38

NOMINAL STAGE CURRENT 1.0 MILLIAMPS
COLLECTOR LOAD RESISTANCE = 12000 OHMS

RBL	RBU	RE	IMA	VG	R(IN)	CG40	CG100
3900	82000	1000	0.90	275	890	28	48
3300	82000	820	0.91	275	860	27	45
3900	100000	680	1.04	295	830	29	49
3900	120000	560	1.00	290	860	29	48
2200	82000	470	0.96	285	740	25	38

NOMINAL STAGE CURRENT 2.0 MILLIAMPS
COLLECTOR LOAD RESISTANCE = 6800 OHMS

RBL	RBU	RE	IMA	VG	R(IN)	CG40	CG100
2200	39000	560	1.91	280	450	29	51
2700	56000	470	1.88	280	480	30	53
2200	56000	390	1.80	275	480	29	49
1800	47000	330	2.05	290	420	29	48
1200	39000	270	1.94	285	380	25	40

NOMINAL STAGE CURRENT 5.0 MILLIAMPS
COLLECTOR LOAD RESISTANCE = 2700 OHMS

RBL	RBU	RE	IMA	VG	R(IN)	CG40	CG100
1500	27000	220	4.58	250	210	32	57
820	18000	180	4.55	250	180	29	49
680	18000	150	4.37	250	180	27	44
1000	27000	120	5.12	265	180	30	52
820	27000	100	4.78	260	180	29	48

NOMINAL STAGE CURRENT 10.0 MILLIAMPS
COLLECTOR LOAD RESISTANCE = 1200 OHMS

RBL	RBU	RE	IMA	VG	R(IN)	CG40	CG100
560	10000	100	10.10	230	90	31	55
560	12000	82	9.87	230	90	31	54
390	10000	68	9.66	230	90	29	48
330	10000	56	9.45	230	90	27	44
330	12000	47	8.70	225	90	27	42

Tables 6 and 7. Transistor Astable and Monostable Circuits

Astable and monostable multivibrators form two invaluable species of circuit for the preparation of a frequency (available in the form of sharp voltage steps) and in the latter case, an output pulse of square shape of a specific length. Each circuit contains two transistors, coupled regeneratively, with appropriate load and bias resistors. The following tables make lists of a range of possible astable and monostable circuits, tabulating their principal characteristic, the basic repetition frequency, or the quasi-stable pulse length, in terms of the collector load resistor (two are used) the base bias resistor, and coupling capacitor(s). The relationship between collector current and maximum base current is chosen so that effective operation will occur with a transistor of current gain of 14 or over, and each of the two complete tables is presented as four separate sets of figures. Each set is defined for a decade sweep of thirteen load resistances (e.g. from 1 kΩ via 1·2 kΩ etc. to 10 kΩ) with appropriate base bias resistances to ensure saturation. This allows a reasonable degree of choice in selecting the operating current for the device depending on the supply voltage used. The average current taken by an astable or monostable multivibrator can be reckoned to be roughly equal to the current taken by one load resistor if it was connected across the supply. In the case of the monostable circuit it is also necessary to define values for the two resistors R_{B2} and R_2 which form the direct coupling between *Tr1* and *Tr2*.

In some cases there may be an additional bias rail available for the return of the 'low' end of R_2, but in order to make construction of the circuits as simple as possible it is assumed that the low end of R_2 will be returned to earth, an arrangement which is satisfactory for most purposes. The value of R_{B2} may be made equal to the quoted value for R_{B1} and R_2 may be selected simply on the nature of the type of transistor being used, e.g. 1·5–3·3 kΩ may be employed for a germanium transistor, and 5·6–10 kΩ for a silicon transistor. There is, however, considerable latitude in the choice of value for R_2 outside these figures.

Operation of the monostable is normally brought about by the application of a positive pulse via C_2 which initiates the cutoff of transistor *Tr1*, prior to regenerative action taking place via *Tr2* and C_1, although initiation may be made by routing a pulse of the opposite polarity to the base of *Tr2*, by a similar network with the diode polarity reversed. The minimum suitable value of capacitor C_2 depends on the magnitude and the risetime of the voltage step which is to be used as a trigger source, but in general suitable capacitances would be in the order of 220 to 1 000 pF, tending towards the lower value when the trigger voltage step is larger. The value of resistor R_3 may be selected on the upper limitation that its time constant with C_2 must be short in relation to the maximum trigger rate that may be applied to the monostable. Practical values may lie between 1 and 10 kΩ.

In the tables, each load resistor is presented in combination with fifteen capacitors (C_1), ranging from 1 000 pF to 15 000 pF, and the corresponding frequency is quoted in hertz, and the length of the monostable pulse is given in milliseconds. For both monostable and astable circuits, the output is directly proportional to the value of C_1 so that extrapolation may be made in both directions outside the fifteen values chosen. Extrapolation is only limited in the short pulse (or high frequency) direction by the high frequency characteristics of the transistors concerned, and in the long pulse (or low frequency) direction by the usual high tolerance of electrolytic capacitors, and the possible presence of leakage currents both across the capacitor and across the back biased transistor. The tables are equally applicable to silicon or germanium transistors, although this will make small differences to the characteristics of the circuit in each case.

In operation, the multivibrator and monostable circuit measure off a time interval which occurs after a charge in C_1 is transferred to the base of *Tr1* causing *Tr1* to be cut off.

The base of *Tr1* is driven into a heavy back bias condition, and C_1 then discharges exponentially via R_{B1} to the supply rail. The time taken for the R_{B1}–C_1 combination to discharge from the initial state (after the instant of switchover) to that potential where *Tr1* just begins to come back into conduction determines the characteristic delay of the monostable, or half the cycle time of the astable circuit (since both sides of the astable function alternately). In both cases, the emitter–base junction of the transistor *Tr1* carries a back bias voltage which is nearly equal to the supply voltage, since this is the magnitude of the voltage change which occurs at the collector of *Tr2* as it switches on, and this potential change is transferred to *Tr1* base by the coupling capacitor. Many transistors however (silicon transistors in particular), are only capable of tolerating a relatively low back bias condition across the emitter–base junction (6–9 V, for example) and in consequence if the supply voltage is greater than this, the transistor junction will break down in the switchover and in consequence will limit the starting voltage of the exponential discharge of the R_{B1}–C_1 combination. Normally, this breakdown does not have a detrimental effect on the transistor but, since it alters the starting point on the exponential decay curve, the result is to reduce the decay time before the reconduction point is reached. Consequently, the operating frequency of a 'breakdown' multivibrator will be greater than that predicted for the non-breakdown case, and the pulse length of a monostable will be shorter than predicted. In order to cater for instances where breakdown may occur, each of the two tables for astable and monostable circuits has been split into four sections, and each section deals with a different degree of breakdown at the emitter–base junction. The first table gives results in that situation where no breakdown occurs at all: the remaining three tables indicate progressively greater levels of emitter–base breakdown, and the tables are placed under heading of a factor 'Discharge Ratio' which represents the ratio of the starting and finishing potentials which occurs across the base bias resistor R_{B1}. In the case where no breakdown occurs the swing of the exponential voltage decay curve at *Tr1* base will change (in the pnp case) from nearly $+2V_{CC}$ with respect to the negative supply rail at the start to approximately $+V_{CC}$ at the finish, giving a Discharge Ratio of $\frac{2V_{CC}}{V_{CC}} = 2$ approximately. Such circuits are therefore tabulated under the heading 'Discharge Ratio = 2·0' in the first table. (In the majority of cases germanium transistors will fall into this category.) As an extreme case of breakdown, consider a monostable running at a supply potential of 24 V using transistors with an emitter–base breakdown of 9 V. The starting potential across R_{B1} will be 24+9, and the potential at which switching will occur will be just less than 24. This gives a Discharge Ratio of $\frac{24+9}{24} = 1{\cdot}38$. The behaviour of such a circuit would correspond with the tables headed with a Discharge Ratio of 1·4.

In cases where the supply voltage exceeds the maximum emitter–base voltage (V_{EB}) of the transistor, therefore, the Discharge Ratio is given by the equation

$$(DR) = \frac{V_{CC}+V_{EB}}{V_{CC}}$$

approximately, and reference should be made to the table with the nearest Discharge Ratio. Some specimen values of the Discharge Ratio for silicon transistors are included in the table.

Since for a given capacitor, the delay of a monostable is directly proportional to the value of R_{B1}, a measure of adjustment can be built into a monostable by making R_{B1} variable in part. The variation should not be allowed to extend to a factor roughly greater than ten since the reduction of R_{B1} causes heavy saturation of *Tr1* and in addition makes the circuit less ready to operate. A variable monostable would then have R_{B1} consisting of at least 10 per cent of the quoted value of R_{B1} as a fixed resistor and the remaining 90 per cent as a variable resistor, and the resulting monostable delay would range from the whole of, to about one-tenth of, the quoted value.

Such a technique does not work well for a multivibrator, since adjustment must be applied to both sides of the circuit to avoid the production of an assymmetric waveform. This can be achieved by returning both base bias resistors to a single decoupled potential divider. Reduction of the base return potential from V_{CC} to $V_{CC}/3$ will approximately halve the operating frequency of the multivibrator.

SPECIMEN VALUES OF DISCHARGE RATIO (DR) FOR SILICON TRANSISTORS

V_{EB}	V_{CC}	(DR)
6	9	1·78
6	12	1·58
6	24	1·28
7	9	1·90
7	12	1·67
7	24	1·32
8	9	2·02
8	12	1·75
8	24	1·37
9	9	2·14
9	12	1·84
9	24	1·41

TABLE. 6. ASTABLE MULTIVIBRATORS.

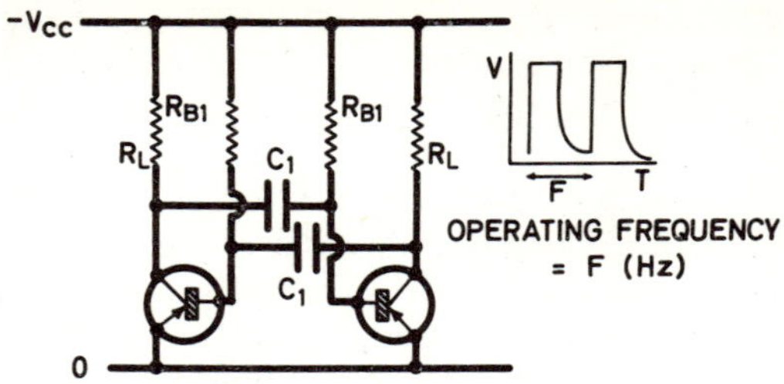

MULTIVIBRATOR CIRCUITS
DISCHARGE RATIO = 2.0

LOAD RESISTANCE 1000 OHMS
BASE RESISTANCE 12000 OHMS

C1(MF)	F(HZ)	C1(MF)	F(HZ)	C1(MF)	F(HZ)
0.0010	60112	0.0012	50094	0.0015	40075
0.0018	33396	0.0022	27324	0.0027	22264
0.0033	18216	0.0039	15413	0.0047	12790
0.0056	10734	0.0068	8840	0.0082	7331
0.0100	6011	0.0120	5009	0.0150	4007

LOAD RESISTANCE 1200 OHMS
BASE RESISTANCE 15000 OHMS

C1(MF)	F(HZ)	C1(MF)	F(HZ)	C1(MF)	F(HZ)
0.0010	48090	0.0012	40075	0.0015	32060
0.0018	26717	0.0022	21859	0.0027	17811
0.0033	14573	0.0039	12331	0.0047	10232
0.0056	8587	0.0068	7072	0.0082	5865
0.0100	4809	0.0120	4007	0.0150	3206

LOAD RESISTANCE 1500 OHMS
BASE RESISTANCE 18000 OHMS

C1(MF)	F(HZ)	C1(MF)	F(HZ)	C1(MF)	F(HZ)
0.0010	40075	0.0012	33396	0.0015	26717
0.0018	22264	0.0022	18216	0.0027	14843
0.0033	12144	0.0039	10276	0.0047	8527
0.0056	7156	0.0068	5893	0.0082	4887
0.0100	4007	0.0120	3340	0.0150	2672

LOAD RESISTANCE 1800 OHMS
BASE RESISTANCE 22000 OHMS

C1(MF)	F(HZ)	C1(MF)	F(HZ)	C1(MF)	F(HZ)
0.0010	32789	0.0012	27324	0.0015	21859
0.0018	18216	0.0022	14904	0.0027	12144
0.0033	9936	0.0039	8407	0.0047	6976
0.0056	5855	0.0068	4822	0.0082	3999
0.0100	3279	0.0120	2732	0.0150	2186

LOAD RESISTANCE 2200 OHMS
BASE RESISTANCE 27000 OHMS

C1(MF)	F(HZ)	C1(MF)	F(HZ)	C1(MF)	F(HZ)
0.0010	26717	0.0012	22264	0.0015	17811
0.0018	14843	0.0022	12144	0.0027	9895
0.0033	8096	0.0039	6850	0.0047	5684
0.0056	4771	0.0068	3929	0.0082	3258
0.0100	2672	0.0120	2226	0.0150	1781

MULTIVIBRATOR CIRCUITS

DISCHARGE RATIO = 2.0

LOAD RESISTANCE 2700 OHMS
BASE RESISTANCE 33000 OHMS

C1(MF)	F(HZ)	C1(MF)	F(HZ)	C1(MF)	F(HZ)
0.0010	21859	0.0012	18216	0.0015	14573
0.0018	12144	0.0022	9936	0.0027	8096
0.0033	6624	0.0039	5605	0.0047	4651
0.0056	3903	0.0068	3215	0.0082	2666
0.0100	2186	0.0120	1822	0.0150	1457

LOAD RESISTANCE 3300 OHMS
BASE RESISTANCE 39000 OHMS

C1(MF)	F(HZ)	C1(MF)	F(HZ)	C1(MF)	F(HZ)
0.0010	18496	0.0012	15413	0.0015	12331
0.0018	10276	0.0022	8407	0.0027	6850
0.0033	5605	0.0039	4743	0.0047	3935
0.0056	3303	0.0068	2720	0.0082	2256
0.0100	1850	0.0120	1541	0.0150	1233

LOAD RESISTANCE 3900 OHMS
BASE RESISTANCE 47000 OHMS

C1(MF)	F(HZ)	C1(MF)	F(HZ)	C1(MF)	F(HZ)
0.0010	15348	0.0012	12790	0.0015	10232
0.0018	8527	0.0022	6976	0.0027	5684
0.0033	4651	0.0039	3935	0.0047	3265
0.0056	2741	0.0068	2257	0.0082	1872
0.0100	1535	0.0120	1279	0.0150	1023

LOAD RESISTANCE 4700 OHMS
BASE RESISTANCE 56000 OHMS

C1(MF)	F(HZ)	C1(MF)	F(HZ)	C1(MF)	F(HZ)
0.0010	12881	0.0012	10734	0.0015	8587
0.0018	7156	0.0022	5855	0.0027	4771
0.0033	3903	0.0039	3303	0.0047	2741
0.0056	2300	0.0068	1894	0.0082	1571
0.0100	1288	0.0120	1073	0.0150	859

LOAD RESISTANCE 5600 OHMS
BASE RESISTANCE 68000 OHMS

C1(MF)	F(HZ)	C1(MF)	F(HZ)	C1(MF)	F(HZ)
0.0010	10608	0.0012	8840	0.0015	7072
0.0018	5893	0.0022	4822	0.0027	3929
0.0033	3215	0.0039	2720	0.0047	2257
0.0056	1894	0.0068	1560	0.0082	1294
0.0100	1061	0.0120	884	0.0150	707

MULTIVIBRATOR CIRCUITS
DISCHARGE RATIO = 2.0→1.8

LOAD RESISTANCE 6800 OHMS
BASE RESISTANCE 82000 OHMS

C1(MF)	F(HZ)	C1(MF)	F(HZ)	C1(MF)	F(HZ)
0.0010	8797	0.0012	7331	0.0015	5865
0.0018	4887	0.0022	3999	0.0027	3258
0.0033	2666	0.0039	2256	0.0047	1872
0.0056	1571	0.0068	1294	0.0082	1073
0.0100	880	0.0120	733	0.0150	586

LOAD RESISTANCE 8200 OHMS
BASE RESISTANCE 100000 OHMS

C1(MF)	F(HZ)	C1(MF)	F(HZ)	C1(MF)	F(HZ)
0.0010	7213	0.0012	6011	0.0015	4809
0.0018	4007	0.0022	3279	0.0027	2672
0.0033	2186	0.0039	1850	0.0047	1535
0.0056	1288	0.0068	1061	0.0082	880
0.0100	721	0.0120	601	0.0150	481

LOAD RESISTANCE 10000 OHMS
BASE RESISTANCE 120000 OHMS

C1(MF)	F(HZ)	C1(MF)	F(HZ)	C1(MF)	F(HZ)
0.0010	6011	0.0012	5009	0.0015	4007
0.0018	3340	0.0022	2732	0.0027	2226
0.0033	1822	0.0039	1541	0.0047	1279
0.0056	1073	0.0068	884	0.0082	733
0.0100	601	0.0120	501	0.0150	401

DISCHARGE RATIO = 1.8
LOAD RESISTANCE 1000 OHMS
BASE RESISTANCE 12000 OHMS

C1(MF)	F(HZ)	C1(MF)	F(HZ)	C1(MF)	F(HZ)
0.0010	70887	0.0012	59073	0.0015	47258
0.0018	39382	0.0022	32222	0.0027	26255
0.0033	21481	0.0039	18176	0.0047	15082
0.0056	12658	0.0068	10425	0.0082	8645
0.0100	7089	0.0120	5907	0.0150	4726

LOAD RESISTANCE 1200 OHMS
BASE RESISTANCE 15000 OHMS

C1(MF)	F(HZ)	C1(MF)	F(HZ)	C1(MF)	F(HZ)
0.0010	56710	0.0012	47258	0.0015	37807
0.0018	31506	0.0022	25777	0.0027	21004
0.0033	17185	0.0039	14541	0.0047	12066
0.0056	10127	0.0068	8340	0.0082	6916
0.0100	5671	0.0120	4726	0.0150	3781

LOAD RESISTANCE 1500 OHMS
BASE RESISTANCE 18000 OHMS

C1(MF)	F(HZ)	C1(MF)	F(HZ)	C1(MF)	F(HZ)
0.0010	47258	0.0012	39382	0.0015	31506
0.0018	26255	0.0022	21481	0.0027	17503
0.0033	14321	0.0039	12118	0.0047	10055
0.0056	8439	0.0068	6950	0.0082	5763
0.0100	4726	0.0120	3938	0.0150	3151

LOAD RESISTANCE 1800 OHMS
BASE RESISTANCE 22000 OHMS

C1(MF)	F(HZ)	C1(MF)	F(HZ)	C1(MF)	F(HZ)
0.0010	38666	0.0012	32222	0.0015	25777
0.0018	21481	0.0022	17575	0.0027	14321
0.0033	11717	0.0039	9914	0.0047	8227
0.0056	6905	0.0068	5686	0.0082	4715
0.0100	3867	0.0120	3222	0.0150	2578

LOAD RESISTANCE 2200 OHMS
BASE RESISTANCE 27000 OHMS

C1(MF)	F(HZ)	C1(MF)	F(HZ)	C1(MF)	F(HZ)
0.0010	31506	0.0012	26255	0.0015	21004
0.0018	17503	0.0022	14321	0.0027	11669
0.0033	9547	0.0039	8078	0.0047	6703
0.0056	5626	0.0068	4633	0.0082	3842
0.0100	3151	0.0120	2625	0.0150	2100

LOAD RESISTANCE 2700 OHMS
BASE RESISTANCE 33000 OHMS

C1(MF)	F(HZ)	C1(MF)	F(HZ)	C1(MF)	F(HZ)
0.0010	25777	0.0012	21481	0.0015	17185
0.0018	14321	0.0022	11717	0.0027	9547
0.0033	7811	0.0039	6610	0.0047	5485
0.0056	4603	0.0068	3791	0.0082	3144
0.0100	2578	0.0120	2148	0.0150	1718

LOAD RESISTANCE 3300 OHMS
BASE RESISTANCE 39000 OHMS

C1(MF)	F(HZ)	C1(MF)	F(HZ)	C1(MF)	F(HZ)
0.0010	21812	0.0012	18176	0.0015	14541
0.0018	12118	0.0022	9914	0.0027	8078
0.0033	6610	0.0039	5593	0.0047	4641
0.0056	3895	0.0068	3208	0.0082	2660
0.0100	2181	0.0120	1818	0.0150	1454

MULTIVIBRATOR CIRCUITS
DISCHARGE RATIO = 1.8

LOAD RESISTANCE 3900 OHMS
BASE RESISTANCE 47000 OHMS

C1(MF)	F(HZ)	C1(MF)	F(HZ)	C1(MF)	F(HZ)
0.0010	18099	0.0012	15082	0.0015	12066
0.0018	10055	0.0022	8227	0.0027	6703
0.0033	5485	0.0039	4641	0.0047	3851
0.0056	3232	0.0068	2662	0.0082	2207
0.0100	1810	0.0120	1508	0.0150	1207

LOAD RESISTANCE 4700 OHMS
BASE RESISTANCE 56000 OHMS

C1(MF)	F(HZ)	C1(MF)	F(HZ)	C1(MF)	F(HZ)
0.0010	15190	0.0012	12658	0.0015	10127
0.0018	8439	0.0022	6905	0.0027	5626
0.0033	4603	0.0039	3895	0.0047	3232
0.0056	2713	0.0068	2234	0.0082	1852
0.0100	1519	0.0120	1266	0.0150	1013

LOAD RESISTANCE 5600 OHMS
BASE RESISTANCE 68000 OHMS

C1(MF)	F(HZ)	C1(MF)	F(HZ)	C1(MF)	F(HZ)
0.0010	12510	0.0012	10425	0.0015	8340
0.0018	6950	0.0022	5686	0.0027	4633
0.0033	3791	0.0039	3208	0.0047	2662
0.0056	2234	0.0068	1840	0.0082	1526
0.0100	1251	0.0120	1042	0.0150	834

LOAD RESISTANCE 6800 OHMS
BASE RESISTANCE 82000 OHMS

C1(MF)	F(HZ)	C1(MF)	F(HZ)	C1(MF)	F(HZ)
0.0010	10374	0.0012	8645	0.0015	6916
0.0018	5763	0.0022	4715	0.0027	3842
0.0033	3144	0.0039	2660	0.0047	2207
0.0056	1852	0.0068	1526	0.0082	1265
0.0100	1037	0.0120	864	0.0150	692

LOAD RESISTANCE 8200 OHMS
BASE RESISTANCE 100000 OHMS

C1(MF)	F(HZ)	C1(MF)	F(HZ)	C1(MF)	F(HZ)
0.0010	8506	0.0012	7089	0.0015	5671
0.0018	4726	0.0022	3867	0.0027	3151
0.0033	2578	0.0039	2181	0.0047	1810
0.0056	1519	0.0068	1251	0.0082	1037
0.0100	851	0.0120	709	0.0150	567

LOAD RESISTANCE 10000 OHMS
BASE RESISTANCE 120000 OHMS

C_1(MF)	F(HZ)	C_1(MF)	F(HZ)	C_1(MF)	F(HZ)
0.0010	7089	0.0012	5907	0.0015	4726
0.0018	3938	0.0022	3222	0.0027	2625
0.0033	2148	0.0039	1818	0.0047	1508
0.0056	1266	0.0068	1042	0.0082	864
0.0100	709	0.0120	591	0.0150	473

DISCHARGE RATIO = 1.6

LOAD RESISTANCE 1000 OHMS
BASE RESISTANCE 12000 OHMS

C_1(MF)	F(HZ)	C_1(MF)	F(HZ)	C_1(MF)	F(HZ)
0.0010	88652	0.0012	73876	0.0015	59101
0.0018	49251	0.0022	40296	0.0027	32834
0.0033	26864	0.0039	22731	0.0047	18862
0.0056	15831	0.0068	13037	0.0082	10811
0.0100	8865	0.0120	7388	0.0150	5910

LOAD RESISTANCE 1200 OHMS
BASE RESISTANCE 15000 OHMS

C_1(MF)	F(HZ)	C_1(MF)	F(HZ)	C_1(MF)	F(HZ)
0.0010	70921	0.0012	59101	0.0015	47281
0.0018	39401	0.0022	32237	0.0027	26267
0.0033	21491	0.0039	18185	0.0047	15090
0.0056	12665	0.0068	10430	0.0082	8649
0.0100	7092	0.0120	5910	0.0150	4728

LOAD RESISTANCE 1500 OHMS
BASE RESISTANCE 18000 OHMS

C_1(MF)	F(HZ)	C_1(MF)	F(HZ)	C_1(MF)	F(HZ)
0.0010	59101	0.0012	49251	0.0015	39401
0.0018	32834	0.0022	26864	0.0027	21889
0.0033	17909	0.0039	15154	0.0047	12575
0.0056	10554	0.0068	8691	0.0082	7207
0.0100	5910	0.0120	4925	0.0150	3940

LOAD RESISTANCE 1800 OHMS
BASE RESISTANCE 22000 OHMS

C_1(MF)	F(HZ)	C_1(MF)	F(HZ)	C_1(MF)	F(HZ)
0.0010	48356	0.0012	40296	0.0015	32237
0.0018	26864	0.0022	21980	0.0027	17909
0.0033	14653	0.0039	12399	0.0047	10288
0.0056	8635	0.0068	7111	0.0082	5897
0.0100	4836	0.0120	4030	0.0150	3224

MULTIVIBRATOR CIRCUITS
DISCHARGE RATIO = 1.6

LOAD RESISTANCE 2200 OHMS
BASE RESISTANCE 27000 OHMS

C1(MF)	F(HZ)	C1(MF)	F(HZ)	C1(MF)	F(HZ)
0.0010	39401	0.0012	32834	0.0015	26267
0.0018	21889	0.0022	17909	0.0027	14593
0.0033	11940	0.0039	10103	0.0047	8383
0.0056	7036	0.0068	5794	0.0082	4805
0.0100	3940	0.0120	3283	0.0150	2627

LOAD RESISTANCE 2700 OHMS
BASE RESISTANCE 33000 OHMS

C1(MF)	F(HZ)	C1(MF)	F(HZ)	C1(MF)	F(HZ)
0.0010	32237	0.0012	26864	0.0015	21491
0.0018	17909	0.0022	14653	0.0027	11940
0.0033	9769	0.0039	8266	0.0047	6859
0.0056	5757	0.0068	4741	0.0082	3931
0.0100	3224	0.0120	2686	0.0150	2149

LOAD RESISTANCE 3300 OHMS
BASE RESISTANCE 39000 OHMS

C1(MF)	F(HZ)	C1(MF)	F(HZ)	C1(MF)	F(HZ)
0.0010	27277	0.0012	22731	0.0015	18185
0.0018	15154	0.0022	12399	0.0027	10103
0.0033	8266	0.0039	6994	0.0047	5804
0.0056	4871	0.0068	4011	0.0082	3327
0.0100	2728	0.0120	2273	0.0150	1818

LOAD RESISTANCE 3900 OHMS
BASE RESISTANCE 47000 OHMS

C1(MF)	F(HZ)	C1(MF)	F(HZ)	C1(MF)	F(HZ)
0.0010	22635	0.0012	18862	0.0015	15090
0.0018	12575	0.0022	10288	0.0027	8383
0.0033	6859	0.0039	5804	0.0047	4816
0.0056	4042	0.0068	3329	0.0082	2760
0.0100	2263	0.0120	1886	0.0150	1509

LOAD RESISTANCE 4700 OHMS
BASE RESISTANCE 56000 OHMS

C1(MF)	F(HZ)	C1(MF)	F(HZ)	C1(MF)	F(HZ)
0.0010	18997	0.0012	15831	0.0015	12665
0.0018	10554	0.0022	8635	0.0027	7036
0.0033	5757	0.0039	4871	0.0047	4042
0.0056	3392	0.0068	2794	0.0082	2317
0.0100	1900	0.0120	1583	0.0150	1266

MULTIVIBRATOR CIRCUITS
DISCHARGE RATIO = 1.6→1.4

LOAD RESISTANCE 5600 OHMS
BASE RESISTANCE 68000 OHMS

C1(MF)	F(HZ)	C1(MF)	F(HZ)	C1(MF)	F(HZ)
0.0010	15644	0.0012	13037	0.0015	10430
0.0018	8691	0.0022	7111	0.0027	5794
0.0033	4741	0.0039	4011	0.0047	3329
0.0056	2794	0.0068	2301	0.0082	1908
0.0100	1564	0.0120	1304	0.0150	1043

LOAD RESISTANCE 6800 OHMS
BASE RESISTANCE 82000 OHMS

C1(MF)	F(HZ)	C1(MF)	F(HZ)	C1(MF)	F(HZ)
0.0010	12973	0.0012	10811	0.0015	8649
0.0018	7207	0.0022	5897	0.0027	4805
0.0033	3931	0.0039	3327	0.0047	2760
0.0056	2317	0.0068	1908	0.0082	1582
0.0100	1297	0.0120	1081	0.0150	865

LOAD RESISTANCE 8200 OHMS
BASE RESISTANCE 100000 OHMS

C1(MF)	F(HZ)	C1(MF)	F(HZ)	C1(MF)	F(HZ)
0.0010	10638	0.0012	8865	0.0015	7092
0.0018	5910	0.0022	4836	0.0027	3940
0.0033	3224	0.0039	2728	0.0047	2263
0.0056	1900	0.0068	1564	0.0082	1297
0.0100	1064	0.0120	887	0.0150	709

LOAD RESISTANCE 10000 OHMS
BASE RESISTANCE 120000 OHMS

C1(MF)	F(HZ)	C1(MF)	F(HZ)	C1(MF)	F(HZ)
0.0010	8865	0.0012	7388	0.0015	5910
0.0018	4925	0.0022	4030	0.0027	3283
0.0033	2686	0.0039	2273	0.0047	1886
0.0056	1583	0.0068	1304	0.0082	1081
0.0100	887	0.0120	739	0.0150	591

DISCHARGE RATIO = 1.4

LOAD RESISTANCE 1000 OHMS
BASE RESISTANCE 12000 OHMS

C1(MF)	F(HZ)	C1(MF)	F(HZ)	C1(MF)	F(HZ)
0.0010	123834	0.0012	103195	0.0015	82556
0.0018	68797	0.0022	56288	0.0027	45864
0.0033	37525	0.0039	31752	0.0047	26348
0.0056	22113	0.0068	18211	0.0082	15102
0.0100	12383	0.0120	10319	0.0150	8256

MULTIVIBRATOR CIRCUITS
DISCHARGE RATIO = 1.4

LOAD RESISTANCE 1200 OHMS
BASE RESISTANCE 15000 OHMS

C1(MF)	F(HZ)	C1(MF)	F(HZ)	C1(MF)	F(HZ)
0.0010	99067	0.0012	82556	0.0015	66045
0.0018	55037	0.0022	45031	0.0027	36692
0.0033	30020	0.0039	25402	0.0047	21078
0.0056	17691	0.0068	14569	0.0082	12081
0.0100	9907	0.0120	8256	0.0150	6604

LOAD RESISTANCE 1500 OHMS
BASE RESISTANCE 18000 OHMS

C1(MF)	F(HZ)	C1(MF)	F(HZ)	C1(MF)	F(HZ)
0.0010	82556	0.0012	68797	0.0015	55037
0.0018	45864	0.0022	37525	0.0027	30576
0.0033	25017	0.0039	21168	0.0047	17565
0.0056	14742	0.0068	12141	0.0082	10068
0.0100	8256	0.0120	6880	0.0150	5504

LOAD RESISTANCE 1800 OHMS
BASE RESISTANCE 22000 OHMS

C1(MF)	F(HZ)	C1(MF)	F(HZ)	C1(MF)	F(HZ)
0.0010	67546	0.0012	56288	0.0015	45031
0.0018	37525	0.0022	30703	0.0027	25017
0.0033	20468	0.0039	17319	0.0047	14371
0.0056	12062	0.0068	9933	0.0082	8237
0.0100	6755	0.0120	5629	0.0150	4503

LOAD RESISTANCE 2200 OHMS
BASE RESISTANCE 27000 OHMS

C1(MF)	F(HZ)	C1(MF)	F(HZ)	C1(MF)	F(HZ)
0.0010	55037	0.0012	45864	0.0015	36692
0.0018	30576	0.0022	25017	0.0027	20384
0.0033	16678	0.0039	14112	0.0047	11710
0.0056	9828	0.0068	8094	0.0082	6712
0.0100	5504	0.0120	4586	0.0150	3669

LOAD RESISTANCE 2700 OHMS
BASE RESISTANCE 33000 OHMS

C1(MF)	F(HZ)	C1(MF)	F(HZ)	C1(MF)	F(HZ)
0.0010	45031	0.0012	37525	0.0015	30020
0.0018	25017	0.0022	20468	0.0027	16678
0.0033	13646	0.0039	11546	0.0047	9581
0.0056	8041	0.0068	6622	0.0082	5492
0.0100	4503	0.0120	3753	0.0150	3002

LOAD RESISTANCE 3300 OHMS
BASE RESISTANCE 39000 OHMS

C1(MF)	F(HZ)	C1(MF)	F(HZ)	C1(MF)	F(HZ)
0.0010	38103	0.0012	31752	0.0015	25402
0.0018	21168	0.0022	17319	0.0027	14112
0.0033	11546	0.0039	9770	0.0047	8107
0.0056	6804	0.0068	5603	0.0082	4647
0.0100	3810	0.0120	3175	0.0150	2540

LOAD RESISTANCE 3900 OHMS
BASE RESISTANCE 47000 OHMS

C1(MF)	F(HZ)	C1(MF)	F(HZ)	C1(MF)	F(HZ)
0.0010	31617	0.0012	26348	0.0015	21078
0.0018	17565	0.0022	14371	0.0027	11710
0.0033	9581	0.0039	8107	0.0047	6727
0.0056	5646	0.0068	4650	0.0082	3856
0.0100	3162	0.0120	2635	0.0150	2108

LOAD RESISTANCE 4700 OHMS
BASE RESISTANCE 56000 OHMS

C1(MF)	F(HZ)	C1(MF)	F(HZ)	C1(MF)	F(HZ)
0.0010	26536	0.0012	22113	0.0015	17691
0.0018	14742	0.0022	12062	0.0027	9828
0.0033	8041	0.0039	6804	0.0047	5646
0.0056	4739	0.0068	3902	0.0082	3236
0.0100	2654	0.0120	2211	0.0150	1769

LOAD RESISTANCE 5600 OHMS
BASE RESISTANCE 68000 OHMS

C1(MF	[illegible]Z)	C1(MF)	F(HZ)	C1(MF)	F(HZ)
0.0[illegible]	[illegible]3	0.0012	18211	0.0015	14569
[illegible]	[illegible]	0.0022	9933	0.0027	8094
[illegible]	[illegible]	0.0039	5603	0.0047	4650
[illegible]	[illegible]	0.0068	3214	0.0082	2665
[illegible]	[illegible]	0.0120	1821	0.0150	1457

[illegible]ISTANCE 6800 OHMS
[illegible]STANCE 82000 OHMS

C1(	[illegible]	[illegible]MF)	F(HZ)	C1(MF)	F(HZ)
0.001[illegible]	[illegible]	[illegible]12	15102	0.0015	12081
0.0018	[illegible]	[illegible]3	8237	0.0027	6712
0.0033	[illegible]	[illegible]	4647	0.0047	3856
0.0056	[illegible]	[illegible]	2665	0.0082	2210
0.0100	[illegible]	[illegible]	1510	0.0150	1208

MULTIVIBRATOR CIRCUITS
DISCHARGE RATIO = 1.4

LOAD RESISTANCE 8200 OHMS
BASE RESISTANCE 100000 OHMS

C1(MF)	F(HZ)	C1(MF)	F(HZ)	C1(MF)	F(HZ)
0.0010	14860	0.0012	12383	0.0015	9907
0.0018	8256	0.0022	6755	0.0027	5504
0.0033	4503	0.0039	3810	0.0047	3162
0.0056	2654	0.0068	2185	0.0082	1812
0.0100	1486	0.0120	1238	0.0150	991

LOAD RESISTANCE 10000 OHMS
BASE RESISTANCE 120000 OHMS

C1(MF)	F(HZ)	C1(MF)	F(HZ)	C1(MF)	F(HZ)
0.0010	12383	0.0012	10319	0.0015	8256
0.0018	6880	0.0022	5629	0.0027	4586
0.0033	3753	0.0039	3175	0.0047	2635
0.0056	2211	0.0068	1821	0.0082	1510
0.0100	1238	0.0120	1032	0.0150	826

SPECIMEN VALUES OF DISCHARGE RATIO (DR) FOR SILICON TRANSISTORS

V_{EB}	V_{CC}	*(DR)*
6	9	1·78
6	12	1·58
6	24	1·28
7	9	1·90
7	12	1·67
7	24	1·32
8	9	2·02
8	12	1·75
8	24	1·37
9	9	2·14
9	12	1·84
9	24	1·41

TABLE. 7. MONOSTABLE MULTIVIBRATORS.

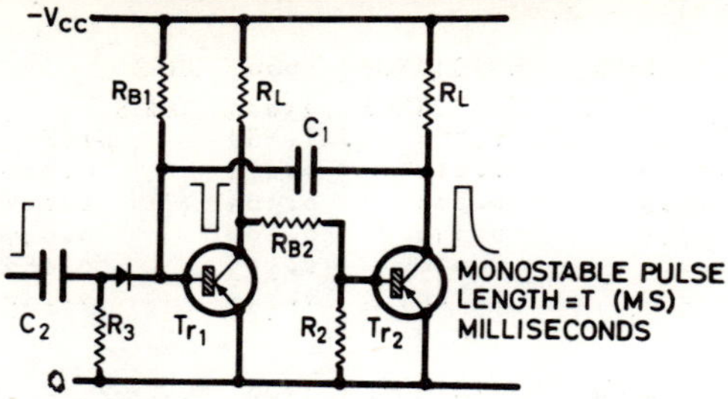

MONOSTABLE CIRCUITS
DISCHARGE RATIO = 2.0

	LOAD	RESISTANCE	1000	OHMS	
	BASE	RESISTANCE	12000	OHMS	
C1(MF)	T(MS)	C1(MF)	T(MS)	C1(MF)	T(MS)
0.010	0.083	0.012	0.100	0.015	0.125
0.018	0.150	0.022	0.183	0.027	0.225
0.033	0.274	0.039	0.324	0.047	0.391
0.056	0.466	0.068	0.566	0.082	0.682
0.100	0.832	0.120	0.998	0.150	1.248

	LOAD	RESISTANCE	1200	OHMS	
	BASE	RESISTANCE	15000	OHMS	
C1(MF)	T(MS)	C1(MF)	T(MS)	C1(MF)	T(MS)
0.010	0.104	0.012	0.125	0.015	0.156
0.018	0.187	0.022	0.229	0.027	0.281
0.033	0.343	0.039	0.405	0.047	0.489
0.056	0.582	0.068	0.707	0.082	0.853
0.100	1.040	0.120	1.248	0.150	1.560

	LOAD	RESISTANCE	1500	OHMS	
	BASE	RESISTANCE	18000	OHMS	
C1(MF)	T(MS)	C1(MF)	T(MS)	C1(MF)	T(MS)
0.010	0.125	0.012	0.150	0.015	0.187
0.018	0.225	0.022	0.274	0.027	0.337
0.033	0.412	0.039	0.487	0.047	0.586
0.056	0.699	0.068	0.848	0.082	1.023
0.100	1.248	0.120	1.497	0.150	1.871

	LOAD	RESISTANCE	1800	OHMS	
	BASE	RESISTANCE	22000	OHMS	
C1(MF)	T(MS)	C1(MF)	T(MS)	C1(MF)	T(MS)
0.010	0.152	0.012	0.183	0.015	0.229
0.018	0.274	0.022	0.335	0.027	0.412
0.033	0.503	0.039	0.595	0.047	0.717
0.056	0.854	0.068	1.037	0.082	1.250
0.100	1.525	0.120	1.830	0.150	2.287

	LOAD	RESISTANCE	2200	OHMS	
	BASE	RESISTANCE	27000	OHMS	
C1(MF)	T(MS)	C1(MF)	T(MS)	C1(MF)	T(MS)
0.010	0.187	0.012	0.225	0.015	0.281
0.018	0.337	0.022	0.412	0.027	0.505
0.033	0.618	0.039	0.730	0.047	0.880
0.056	1.048	0.068	1.273	0.082	1.535
0.100	1.871	0.120	2.246	0.150	2.807

MONOSTABLE CIRCUITS
DISCHARGE RATIO = 2.0

LOAD RESISTANCE 2700 OHMS
BASE RESISTANCE 33000 OHMS

C1(MF)	T(MS)	C1(MF)	T(MS)	C1(MF)	T(MS)
0.010	0.229	0.012	0.274	0.015	0.343
0.018	0.412	0.022	0.503	0.027	0.618
0.033	0.755	0.039	0.892	0.047	1.075
0.056	1.281	0.068	1.555	0.082	1.876
0.100	2.287	0.120	2.745	0.150	3.431

LOAD RESISTANCE 3300 OHMS
BASE RESISTANCE 39000 OHMS

C1(MF)	T(MS)	C1(MF)	T(MS)	C1(MF)	T(MS)
0.010	0.270	0.012	0.324	0.015	0.405
0.018	0.487	0.022	0.595	0.027	0.730
0.033	0.892	0.039	1.054	0.047	1.271
0.056	1.514	0.068	1.838	0.082	2.217
0.100	2.703	0.120	3.244	0.150	4.055

LOAD RESISTANCE 3900 OHMS
BASE RESISTANCE 47000 OHMS

C1(MF)	T(MS)	C1(MF)	T(MS)	C1(MF)	T(MS)
0.010	0.326	0.012	0.391	0.015	0.489
0.018	0.586	0.022	0.717	0.027	0.880
0.033	1.075	0.039	1.271	0.047	1.531
0.056	1.824	0.068	2.215	0.082	2.671
0.100	3.258	0.120	3.909	0.150	4.887

LOAD RESISTANCE 4700 OHMS
BASE RESISTANCE 56000 OHMS

C1(MF)	T(MS)	C1(MF)	T(MS)	C1(MF)	T(MS)
0.010	0.388	0.012	0.466	0.015	0.582
0.018	0.699	0.022	0.854	0.027	1.048
0.033	1.281	0.039	1.514	0.047	1.824
0.056	2.174	0.068	2.640	0.082	3.183
0.100	3.882	0.120	4.658	0.150	5.822

LOAD RESISTANCE 5600 OHMS
BASE RESISTANCE 68000 OHMS

C1(MF)	T(MS)	C1(MF)	T(MS)	C1(MF)	T(MS)
0.010	0.471	0.012	0.566	0.015	0.707
0.018	0.848	0.022	1.037	0.027	1.273
0.033	1.555	0.039	1.838	0.047	2.215
0.056	2.640	0.068	3.205	0.082	3.865
0.100	4.713	0.120	5.656	0.150	7.070

LOAD RESISTANCE 6800 OHMS
BASE RESISTANCE 82000 OHMS

C1(MF)	T(MS)	C1(MF)	T(MS)	C1(MF)	T(MS)
0.010	0.568	0.012	0.682	0.015	0.853
0.018	1.023	0.022	1.250	0.027	1.535
0.033	1.876	0.039	2.217	0.047	2.671
0.056	3.183	0.068	3.865	0.082	4.661
0.100	5.684	0.120	6.821	0.150	8.526

LOAD RESISTANCE 8200 OHMS
BASE RESISTANCE 100000 OHMS

C1(MF)	T(MS)	C1(MF)	T(MS)	C1(MF)	T(MS)
0.010	0.693	0.012	0.832	0.015	1.040
0.018	1.248	0.022	1.525	0.027	1.871
0.033	2.287	0.039	2.703	0.047	3.258
0.056	3.882	0.068	4.713	0.082	5.684
0.100	6.931	0.120	8.318	0.150	10.397

LOAD RESISTANCE 10000 OHMS
BASE RESISTANCE 120000 OHMS

C1(MF)	T(MS)	C1(MF)	T(MS)	C1(MF)	T(MS)
0.010	0.832	0.012	0.998	0.015	1.248
0.018	1.497	0.022	1.830	0.027	2.246
0.033	2.745	0.039	3.244	0.047	3.909
0.056	4.658	0.068	5.656	0.082	6.821
0.100	8.318	0.120	9.981	0.150	12.477

DISCHARGE RATIO = 1.8

LOAD RESISTANCE 1000 OHMS
BASE RESISTANCE 12000 OHMS

C1(MF)	T(MS)	C1(MF)	T(MS)	C1(MF)	T(MS)
0.010	0.071	0.012	0.085	0.015	0.106
0.018	0.127	0.022	0.155	0.027	0.190
0.033	0.233	0.039	0.275	0.047	0.332
0.056	0.395	0.068	0.480	0.082	0.578
0.100	0.705	0.120	0.846	0.150	1.058

LOAD RESISTANCE 1200 OHMS
BASE RESISTANCE 15000 OHMS

C1(MF)	T(MS)	C1(MF)	T(MS)	C1(MF)	T(MS)
0.010	0.088	0.012	0.106	0.015	0.132
0.018	0.159	0.022	0.194	0.027	0.238
0.033	0.291	0.039	0.344	0.047	0.414
0.056	0.494	0.068	0.600	0.082	0.723
0.100	0.882	0.120	1.058	0.150	1.323

MONOSTABLE CIRCUITS

DISCHARGE RATIO = 1.8

LOAD RESISTANCE 1500 OHMS
BASE RESISTANCE 18000 OHMS

C1(MF)	T(MS)	C1(MF)	T(MS)	C1(MF)	T(MS)
0.010	0.106	0.012	0.127	0.015	0.159
0.018	0.190	0.022	0.233	0.027	0.286
0.033	0.349	0.039	0.413	0.047	0.497
0.056	0.592	0.068	0.719	0.082	0.868
0.100	1.058	0.120	1.270	0.150	1.587

LOAD RESISTANCE 1800 OHMS
BASE RESISTANCE 22000 OHMS

C1(MF)	T(MS)	C1(MF)	T(MS)	C1(MF)	T(MS)
0.010	0.129	0.012	0.155	0.015	0.194
0.018	0.233	0.022	0.284	0.027	0.349
0.033	0.427	0.039	0.504	0.047	0.608
0.056	0.724	0.068	0.879	0.082	1.060
0.100	1.293	0.120	1.552	0.150	1.940

LOAD RESISTANCE 2200 OHMS
BASE RESISTANCE 27000 OHMS

C1(MF)	T(MS)	C1(MF)	T(MS)	C1(MF)	T(MS)
0.010	0.159	0.012	0.190	0.015	0.238
0.018	0.286	0.022	0.349	0.027	0.428
0.033	0.524	0.039	0.619	0.047	0.746
0.056	0.889	0.068	1.079	0.082	1.301
0.100	1.587	0.120	1.904	0.150	2.381

LOAD RESISTANCE 2700 OHMS
BASE RESISTANCE 33000 OHMS

C1(MF)	T(MS)	C1(MF)	T(MS)	C1(MF)	T(MS)
0.010	0.194	0.012	0.233	0.015	0.291
0.018	0.349	0.022	0.427	0.027	0.524
0.033	0.640	0.039	0.756	0.047	0.912
0.056	1.086	0.068	1.319	0.082	1.591
0.100	1.940	0.120	2.328	0.150	2.910

LOAD RESISTANCE 3300 OHMS
BASE RESISTANCE 39000 OHMS

C1(MF)	T(MS)	C1(MF)	T(MS)	C1(MF)	T(MS)
0.010	0.229	0.012	0.275	0.015	0.344
0.018	0.413	0.022	0.504	0.027	0.619
0.033	0.756	0.039	0.894	0.047	1.077
0.056	1.284	0.068	1.559	0.082	1.880
0.100	2.292	0.120	2.751	0.150	3.439

MONOSTABLE CIRCUITS

DISCHARGE RATIO = 1.8

LOAD RESISTANCE 3900 OHMS
BASE RESISTANCE 47000 OHMS

C1(MF)	T(MS)	C1(MF)	T(MS)	C1(MF)	T(MS)
0.010	0.276	0.012	0.332	0.015	0.414
0.018	0.497	0.022	0.608	0.027	0.746
0.033	0.912	0.039	1.077	0.047	1.298
0.056	1.547	0.068	1.879	0.082	2.265
0.100	2.763	0.120	3.315	0.150	4.144

LOAD RESISTANCE 4700 OHMS
BASE RESISTANCE 56000 OHMS

C1(MF)	T(MS)	C1(MF)	T(MS)	C1(MF)	T(MS)
0.010	0.329	0.012	0.395	0.015	0.494
0.018	0.592	0.022	0.724	0.027	0.889
0.033	1.086	0.039	1.284	0.047	1.547
0.056	1.843	0.068	2.238	0.082	2.699
0.100	3.292	0.120	3.950	0.150	4.937

LOAD RESISTANCE 5600 OHMS
BASE RESISTANCE 68000 OHMS

C1(MF)	T(MS)	C1(MF)	T(MS)	C1(MF)	T(MS)
0.010	0.400	0.012	0.480	0.015	0.600
0.018	0.719	0.022	0.879	0.027	1.079
0.033	1.319	0.039	1.559	0.047	1.879
0.056	2.238	0.068	2.718	0.082	3.277
0.100	3.997	0.120	4.796	0.150	5.995

LOAD RESISTANCE 6800 OHMS
BASE RESISTANCE 82000 OHMS

C1(MF)	T(MS)	C1(MF)	T(MS)	C1(MF)	T(MS)
0.010	0.482	0.012	0.578	0.015	0.723
0.018	0.868	0.022	1.060	0.027	1.301
0.033	1.591	0.039	1.880	0.047	2.265
0.056	2.699	0.068	3.277	0.082	3.952
0.100	4.820	0.120	5.784	0.150	7.230

LOAD RESISTANCE 8200 OHMS
BASE RESISTANCE 100000 OHMS

C1(MF)	T(MS)	C1(MF)	T(MS)	C1(MF)	T(MS)
0.010	0.588	0.012	0.705	0.015	0.882
0.018	1.058	0.022	1.293	0.027	1.587
0.033	1.940	0.039	2.292	0.047	2.763
0.056	3.292	0.068	3.997	0.082	4.820
0.100	5.878	0.120	7.053	0.150	8.817

MONOSTABLE CIRCUITS

DISCHARGE RATIO = 1.8→1.6

LOAD RESISTANCE 10000 OHMS
BASE RESISTANCE 120000 OHMS

C1(MF)	T(MS)	C1(MF)	T(MS)	C1(MF)	T(MS)
0.010	0.705	0.012	0.846	0.015	1.058
0.018	1.270	0.022	1.552	0.027	1.904
0.033	2.328	0.039	2.751	0.047	3.315
0.056	3.950	0.068	4.796	0.082	5.784
0.100	7.053	0.120	8.464	0.150	10.580

DISCHARGE RATIO = 1.6

LOAD RESISTANCE 1000 OHMS
BASE RESISTANCE 12000 OHMS

C1(MF)	T(MS)	C1(MF)	T(MS)	C1(MF)	T(MS)
0.010	0.056	0.012	0.068	0.015	0.085
0.018	0.102	0.022	0.124	0.027	0.152
0.033	0.186	0.039	0.220	0.047	0.265
0.056	0.316	0.068	0.384	0.082	0.462
0.100	0.564	0.120	0.677	0.150	0.846

LOAD RESISTANCE 1200 OHMS
BASE RESISTANCE 15000 OHMS

C1(MF)	T(MS)	C1(MF)	T(MS)	C1(MF)	T(MS)
0.010	0.071	0.012	0.085	0.015	0.106
0.018	0.127	0.022	0.155	0.027	0.190
0.033	0.233	0.039	0.275	0.047	0.331
0.056	0.395	0.068	0.479	0.082	0.578
0.100	0.705	0.120	0.846	0.150	1.058

LOAD RESISTANCE 1500 OHMS
BASE RESISTANCE 18000 OHMS

C1(MF)	T(MS)	C1(MF)	T(MS)	C1(MF)	T(MS)
0.010	0.085	0.012	0.102	0.015	0.127
0.018	0.152	0.022	0.186	0.027	0.228
0.033	0.279	0.039	0.330	0.047	0.398
0.056	0.474	0.068	0.575	0.082	0.694
0.100	0.846	0.120	1.015	0.150	1.269

LOAD RESISTANCE 1800 OHMS
BASE RESISTANCE 22000 OHMS

C1(MF)	T(MS)	C1(MF)	T(MS)	C1(MF)	T(MS)
0.010	0.103	0.012	0.124	0.015	0.155
0.018	0.186	0.022	0.227	0.027	0.279
0.033	0.341	0.039	0.403	0.047	0.486
0.056	0.579	0.068	0.703	0.082	0.848
0.100	1.034	0.120	1.241	0.150	1.551

LOAD RESISTANCE 2200 OHMS
BASE RESISTANCE 27000 OHMS

C1(MF)	T(MS)	C1(MF)	T(MS)	C1(MF)	T(MS)
0.010	0.127	0.012	0.152	0.015	0.190
0.018	0.228	0.022	0.279	0.027	0.343
0.033	0.419	0.039	0.495	0.047	0.596
0.056	0.711	0.068	0.863	0.082	1.041
0.100	1.269	0.120	1.523	0.150	1.904

LOAD RESISTANCE 2700 OHMS
BASE RESISTANCE 33000 OHMS

C1(MF)	T(MS)	C1(MF)	T(MS)	C1(MF)	T(MS)
0.010	0.155	0.012	0.186	0.015	0.233
0.018	0.279	0.022	0.341	0.027	0.419
0.033	0.512	0.039	0.605	0.047	0.729
0.056	0.869	0.068	1.055	0.082	1.272
0.100	1.551	0.120	1.861	0.150	2.327

LOAD RESISTANCE 3300 OHMS
BASE RESISTANCE 39000 OHMS

C1(MF)	T(MS)	C1(MF)	T(MF)	C1(MF)	T(MS)
0.010	0.183	0.012	0.220	0.015	0.275
0.018	0.330	0.022	0.403	0.027	0.495
0.033	0.605	0.039	0.715	0.047	0.862
0.056	1.026	0.068	1.246	0.082	1.503
0.100	1.833	0.120	2.200	0.150	2.750

LOAD RESISTANCE 3900 OHMS
BASE RESISTANCE 47000 OHMS

C1(MF)	T(MS)	C1(MF)	T(MS)	C1(MF)	T(MS)
0.010	0.221	0.012	0.265	0.015	0.331
0.018	0.398	0.022	0.486	0.027	0.596
0.033	0.729	0.039	0.862	0.047	1.038
0.056	1.237	0.068	1.502	0.082	1.811
0.100	2.209	0.120	2.651	0.150	3.314

LOAD RESISTANCE 4700 OHMS
BASE RESISTANCE 56000 OHMS

C1(MF)	T(MS)	C1(MF)	T(MS)	C1(MF)	T(MS)
0.010	0.263	0.012	0.316	0.015	0.395
0.018	0.474	0.022	0.579	0.027	0.711
0.033	0.869	0.039	1.026	0.047	1.237
0.056	1.474	0.068	1.790	0.082	2.158
0.100	2.632	0.120	3.158	0.150	3.948

MONOSTABLE CIRCUITS

DISCHARGE RATIO = 1.6→1.4

LOAD RESISTANCE 5600 OHMS
BASE RESISTANCE 68000 OHMS

C1(MF)	T(MS)	C1(MF)	T(MS)	C1(MF)	T(MS)
0.010	0.320	0.012	0.384	0.015	0.479
0.018	0.575	0.022	0.703	0.027	0.863
0.033	1.055	0.039	1.246	0.047	1.502
0.056	1.790	0.068	2.173	0.082	2.621
0.100	3.196	0.120	3.835	0.150	4.794

LOAD RESISTANCE 6800 OHMS
BASE RESISTANCE 82000 OHMS

C1(MF)	T(MS)	C1(MF)	T(MS)	C1(MF)	T(MS)
0.010	0.385	0.012	0.462	0.015	0.578
0.018	0.694	0.022	0.848	0.027	1.041
0.033	1.272	0.039	1.503	0.047	1.811
0.056	2.158	0.068	2.621	0.082	3.160
0.100	3.854	0.120	4.625	0.150	5.781

LOAD RESISTANCE 8200 OHMS
BASE RESISTANCE 100000 OHMS

C1(MF)	T(MS)	C1(MF)	T(MS)	C1(MF)	T(MS)
0.010	0.470	0.012	0.564	0.015	0.705
0.018	0.846	0.022	1.034	0.027	1.269
0.033	1.551	0.039	1.833	0.047	2.209
0.056	2.632	0.068	3.196	0.082	3.854
0.100	4.700	0.120	5.640	0.150	7.050

LOAD RESISTANCE 10000 OHMS
BASE RESISTANCE 120000 OHMS

C1(MF)	T(MS)	C1(MF)	T(MS)	C1(MF)	T(MS)
0.010	0.564	0.012	0.677	0.015	0.846
0.018	1.015	0.022	1.241	0.027	1.523
0.033	1.861	0.039	2.200	0.047	2.651
0.056	3.158	0.068	3.835	0.082	4.625
0.100	5.640	0.120	6.768	0.150	8.460

DISCHARGE RATIO = 1.4

LOAD RESISTANCE 1000 OHMS
BASE RESISTANCE 12000 OHMS

C1(MF)	T(MS)	C1(MF)	T(MS)	C1(MF)	T(MS)
0.010	0.040	0.012	0.048	0.015	0.061
0.018	0.073	0.022	0.089	0.027	0.109
0.033	0.133	0.039	0.157	0.047	0.190
0.056	0.226	0.068	0.275	0.082	0.331
0.100	0.404	0.120	0.485	0.150	0.606

DISCHARGE RATIO = 1.4

LOAD RESISTANCE 1200 OHMS
BASE RESISTANCE 15000 OHMS

C1(MF)	T(MS)	C1(MF)	T(MS)	C1(MF)	T(MS)
0.010	0.050	0.012	0.061	0.015	0.076
0.018	0.091	0.022	0.111	0.027	0.136
0.033	0.167	0.039	0.197	0.047	0.237
0.056	0.283	0.068	0.343	0.082	0.414
0.100	0.505	0.120	0.606	0.150	0.757

LOAD RESISTANCE 1500 OHMS
BASE RESISTANCE 18000 OHMS

C1(MF)	T(MS)	C1(MF)	T(MS)	C1(MF)	T(MS)
0.010	0.061	0.012	0.073	0.015	0.091
0.018	0.109	0.022	0.133	0.027	0.164
0.033	0.200	0.039	0.236	0.047	0.285
0.056	0.339	0.068	0.412	0.082	0.497
0.100	0.606	0.120	0.727	0.150	0.908

LOAD RESISTANCE 1800 OHMS
BASE RESISTANCE 22000 OHMS

C1(MF)	T(MS)	C1(MF)	T(MS)	C1(MF)	T(MS)
0.010	0.074	0.012	0.089	0.015	0.111
0.018	0.133	0.022	0.163	0.027	0.200
0.033	0.244	0.039	0.289	0.047	0.348
0.056	0.415	0.068	0.503	0.082	0.607
0.100	0.740	0.120	0.888	0.150	1.110

LOAD RESISTANCE 2200 OHMS
BASE RESISTANCE 27000 OHMS

C1(MF)	T(MS)	C1(MF)	T(MS)	C1(MF)	T(MS)
0.010	0.091	0.012	0.109	0.015	0.136
0.018	0.164	0.022	0.200	0.027	0.245
0.033	0.300	0.039	0.354	0.047	0.427
0.056	0.509	0.068	0.618	0.082	0.745
0.100	0.908	0.120	1.090	0.150	1.363

LOAD RESISTANCE 2700 OHMS
BASE RESISTANCE 33000 OHMS

C1(MF)	T(MS)	C1(MF)	T(MS)	C1(MF)	T(MS)
0.010	0.111	0.012	0.133	0.015	0.167
0.018	0.200	0.022	0.244	0.027	0.300
0.033	0.366	0.039	0.433	0.047	0.522
0.056	0.622	0.068	0.755	0.082	0.910
0.100	1.110	0.120	1.332	0.150	1.666

MONOSTABLE CIRCUITS
DISCHARGE RATIO = 1.4

LOAD RESISTANCE 3300 OHMS
BASE RESISTANCE 39000 OHMS

C1(MF)	T(MS)	C1(MF)	T(MS)	C1(MF)	T(MS)
0.010	0.131	0.012	0.157	0.015	0.197
0.018	0.236	0.022	0.289	0.027	0.354
0.033	0.433	0.039	0.512	0.047	0.617
0.056	0.735	0.068	0.892	0.082	1.076
0.100	1.312	0.120	1.575	0.150	1.968

LOAD RESISTANCE 3900 OHMS
BASE RESISTANCE 47000 OHMS

C1(MF)	T(MS)	C1(MF)	T(MS)	C1(MF)	T(MS)
0.010	0.158	0.012	0.190	0.015	0.237
0.018	0.285	0.022	0.348	0.027	0.427
0.033	0.522	0.039	0.617	0.047	0.743
0.056	0.886	0.068	1.075	0.082	1.297
0.100	1.581	0.120	1.898	0.150	2.372

LOAD RESISTANCE 4700 OHMS
BASE RESISTANCE 56000 OHMS

C1(MF)	T(MS)	C1(MF)	T(MS)	C1(MF)	T(MS)
0.010	0.188	0.012	0.226	0.015	0.283
0.018	0.339	0.022	0.415	0.027	0.509
0.033	0.622	0.039	0.735	0.047	0.886
0.056	1.055	0.068	1.281	0.082	1.545
0.100	1.884	0.120	2.261	0.150	2.826

LOAD RESISTANCE 5600 OHMS
BASE RESISTANCE 68000 OHMS

C1(MF)	T(MS)	C1(MF)	T(MS)	C1(MF)	T(MS)
0.010	0.229	0.012	0.275	0.015	0.343
0.018	0.412	0.022	0.503	0.027	0.618
0.033	0.755	0.039	0.892	0.047	1.075
0.056	1.281	0.068	1.556	0.082	1.876
0.100	2.288	0.120	2.746	0.150	3.432

LOAD RESISTANCE 6800 OHMS
BASE RESISTANCE 82000 OHMS

C1(MF)	T(MS)	C1(MF)	T(MS)	C1(MF)	T(MS)
0.010	0.276	0.012	0.331	0.015	0.414
0.018	0.497	0.022	0.607	0.027	0.745
0.033	0.910	0.039	1.076	0.047	1.297
0.056	1.545	0.068	1.876	0.082	2.262
0.100	2.759	0.120	3.311	0.150	4.139

LOAD RESISTANCE 8200 OHMS
BASE RESISTANCE 100000 OHMS

C1(MF)	T(MS)	C1(MF)	T(MS)	C1(MF)	T(MS)
0.010	0.336	0.012	0.404	0.015	0.505
0.018	0.606	0.022	0.740	0.027	0.908
0.033	1.110	0.039	1.312	0.047	1.581
0.056	1.884	0.068	2.288	0.082	2.759
0.100	3.365	0.120	4.038	0.150	5.047

LOAD RESISTANCE 10000 OHMS
BASE RESISTANCE 120000 OHMS

C1(MF)	T(MS)	C1(MF)	T(MS)	C1(MF)	T(MS)
0.010	0.404	0.012	0.485	0.015	0.606
0.018	0.727	0.022	0.888	0.027	1.090
0.033	1.332	0.039	1.575	0.047	1.898
0.056	2.261	0.068	2.746	0.082	3.311
0.100	4.038	0.120	4.845	0.150	6.057

Table 8. Schmitt Trigger Circuits

Of all the circuits described in this book, the Schmitt Trigger is invariably the one which gives the greatest of difficulty from the practical designer's point of view. In its basic form, the Schmitt Trigger consists of two transistors (*Tr1* and *Tr2*) linked by a common emitter resistor (R_E), with the output of the first stage (*Tr2*) fed directly by potential divider (R_1 and R_2), to the input of the second (*Tr2*). In the absence of base drive to *Tr1*, that transistor is cut off and a current passes through *Tr2* collector via R_E, the potential at *Tr2* base being established by (R_L and R_1) and R_2 acting as a potential divider. If the base of *Tr1* is now connected to a slowly negative going potential (considering the pnp case), *Tr1* will begin to conduct when its base is just negative with respect to its emitter, which is held at a specific potential by the action of *Tr2* as described previously. Conduction of *Tr1* brings the potential of the junction of R_L and R_1 more positive, resulting in a decrease of current through *Tr2*. Regenerative switching occurs, with the attainment of the ultimate situation where *Tr1* saturates and *Tr2* is cut off by a state of back bias applied by the action of R_1 and R_2. The circuit remains in this state until the situation is reversed as the potential of *Tr1* base is once again brought more positive. *Tr1* comes out of saturation and the circuit regenerates again when the negative going potential of the R_1/R_2 junction and Tr_2 base crosses the positive going potential of *Tr2* emitter, since this defines the instant when *Tr2* starts to conduct again.

The consequence of this behaviour is that the trigger circuit will switch states (observable as either a state of cutoff or as the passage of a specific collector current in *Tr2*) at two specific voltage levels corresponding to the two specific threshold conditions described above. These voltage levels are defined as the 'restore voltage' (*VR*) being that potential where *Tr1* passes into conduction, and the 'trip voltage' (*VT*) where *Tr2* conducts, and form the two most important properties of the Schmitt Trigger circuit. The third significant property is of course the magnitude of the current which is available from *Tr2* collector. The interdependence of the two trip levels on the four resistances making up the circuit is such that the whole circuit defies a simple design technique and therefore can prove very troublesome to the experimenter, who normally is interested in designing for a specific pair of trip levels to give a definite 'backlash' (i.e. difference between restore and trip voltage) in any situation.

The basic mathematics underlying the operation of a Schmitt Trigger circuit has been described in 'Wireless World', for example, where in March and April, 1967, A. E. Crump gave a very comprehensive theoretical treatment of the subject.

The following table has been compiled with the aid of the computer to give nearly thirteen hundred Schmitt Trigger circuits to choose from, defining the necessary values of the four relevant resistors to give a specific trip and restore voltage. Computations were made in part by a trial-and-error technique which was necessary because of the restriction of having only preferred resistance values available. Printout was withheld in all cases where the desired trip and restore voltages could not be met closely, so the continuity of the tables is rather sporadic as a result. The tables are evenly divided between silicon and germanium transistors, and then into 160 examples at each of the four supply voltages used elsewhere in the tables—6, 9, 12, and 24 V. The increments of voltage in each case have been varied in order to optimise the coverage in the space available: for example, 6 and 9 V circuits ascend in steps of restore voltage (the larger of the pair) of 0·2 V, 12 V circuits, steps of 0·25 V, and 24 V circuits in steps of 0·5 V.

These allow the table to ascend to give trigger voltages extending to well over half the available supply voltage: it is even feasible to obtain higher operating levels than this by changing the polarity of the transistors—for example a pnp Schmitt required to operate at 21·0 and 20·0 V on a 24 V supply could be obtained by using a complementary 4·0/3·0 V npn version, although the output current would have the opposite polarity and sense.*

In all cases, the available current at *Tr2* collector has been designed initially to be 1 mA. No value of resistance has been quoted for the collector load of *Tr2*, since it is left to the designer to decide how he wishes to make use of this current. The two logical alternatives are to apply it as base drive to a complementary transistor, or if a voltage wave is required, to insert a suitable resistance to convert the current into a square voltage of appropriate height. In no case should the load resistance applied to *Tr2* be so large as to cause saturation since this will affect the operating levels.

If it is necessary to produce a trigger in which a larger output current is available from *Tr2* collector (such as currents of 10 or even 100 mA), suitable circuits can be deduced from the tables simply by dividing each of the four quoted resistance values by the same factor of 10 or 100. The effect on the trip and restore voltages is small in most cases. This would, of course, allow the tabulated trigger circuits to be used for switching loads such as relays or lamps.

When silicon transistors are being used for a Schmitt Trigger, it

* (The circuits may, of course, be used in the npn as well as the pnp configuration).

is sometimes possible to exceed the emitter–base breakdown voltage (V_{BE}) of *Tr1* under conditions of large drive voltage during the condition where *Tr1* is cut off. As a result, additional current is injected into the emitter circuit of *Tr2*, which increases the output current from the trigger. This undesirable effect may be overcome by the use of an additional diode in the base of *Tr1*, or preferably, by selection of germanium transistors for such applications.

For the basic sine-to-square wave conversion, a trigger circuit may be chosen in conjunction with a potential divider both for selection of the minimum threshold at which the trigger will operate, and for optimisation of the mark-space ratio of the output current. The trigger backlash (i.e. $V_R - V_T$), is chosen to be equal to the minimum peak-to-peak value of the incoming wave above which squaring is to take place, and the quiescent potential of the potential divider is selected to be near $(V_R + V_T)/2$, i.e. midway between the upper and lower trip voltages. This ensures that as soon as the circuit starts to function, the resulting output wave will have a mark-space ratio near unity. The signal is injected by capacitor into the potential divider-trigger combination.

A criterion of the performance of the Schmitt Trigger is termed the 'loop gain' of the circuit. Broadly speaking, the circuit will only act regeneratively, that is to say, it will only switch abruptly between its two extreme states, if its gain is greater than unity. If the loop gain of a Schmitt Trigger is near unity the circuit is liable to 'slide' between its two states, and can be held in any intermediate position by suitable adjustment of the input voltage. In order to make the calculation for loop gain it is necessary to assume a value for the internal resistance of the voltage source to which the trigger is connected, and for each trigger circuit a value for loop gain (*LG*) has been quoted on the assumption that the source resistance is 2000 Ω. As the source resistance increases, the loop gain of the circuit decreases, so the loop gain values are presented as a guide for selection, since if a trigger is required to operate satisfactorily from a high resistance voltage source, a trigger combination with a high loop gain should be chosen in preference.

TABLE. 8. SCHMITT TRIGGER CIRCUITS.

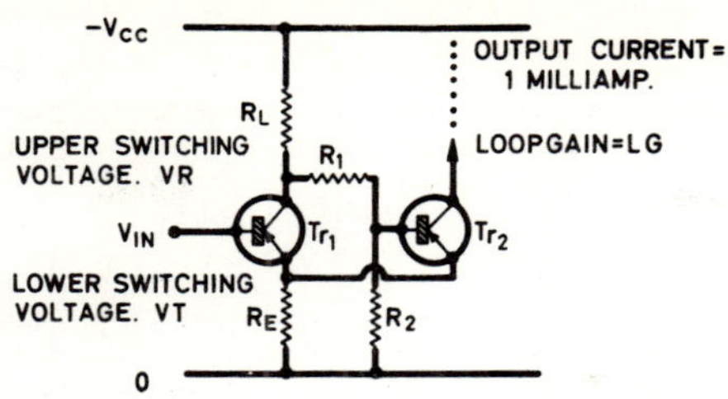

SCHMITT TRIGGER CIRCUITS

SILICON 6 VOLT

VR	VT	RE	R2	RL	R1	LG
1.0	0.8	270	3900	5600	10000	5.2
1.2	0.8	470	4700	15000	820	15.9
1.2	0.9	470	4700	5600	10000	6.0
1.2	1.0	470	4700	3300	12000	3.6
1.2	1.1	470	4700	1500	15000	1.5
1.4	0.8	680	5600	15000	680	17.6
1.4	0.9	680	5600	10000	5600	11.8
1.4	1.0	680	5600	5600	10000	6.6
1.4	1.1	680	5600	3900	12000	4.5
1.4	1.3	680	5600	1200	15000	1.4
1.6	0.9	820	6800	15000	820	18.9
1.6	1.0	820	6800	8200	8200	10.0
1.6	1.1	820	6800	5600	10000	7.1
1.6	1.2	820	6800	3900	12000	4.9
1.6	1.4	820	6800	1500	15000	1.8
1.6	1.5	820	6800	1200	15000	1.5
1.8	0.9	1000	8200	15000	1200	19.8
1.8	1.0	1000	8200	12000	3900	16.1
1.8	1.1	1000	8200	8200	8200	10.7
1.8	1.2	1000	8200	5600	10000	7.6
1.8	1.3	1000	8200	3900	12000	5.2
1.8	1.5	1000	8200	1800	15000	2.3
1.8	1.6	1000	8200	1500	15000	1.9
1.8	1.7	1000	8200	1200	15000	1.6
2.0	1.0	1200	8200	12000	2200	17.8
2.0	1.1	1200	8200	10000	3900	15.1
2.0	1.2	1200	8200	8200	5600	12.5
2.0	1.3	1200	8200	5600	8200	8.5
2.0	1.4	1200	8200	3900	10000	5.9
2.0	1.6	1200	8200	2200	12000	3.3
2.0	1.7	1200	8200	1800	12000	2.7
2.2	1.1	1200	10000	10000	4700	15.2
2.2	1.4	1200	10000	4700	10000	7.1
2.2	1.6	1200	10000	2700	12000	4.1
2.2	1.8	1200	10000	2200	12000	3.4
2.4	1.2	1500	10000	10000	2700	17.3
2.4	1.3	1500	10000	8200	4700	14.0
2.4	1.6	1500	10000	4700	8200	8.0
2.4	1.7	1500	10000	3300	10000	5.5
2.4	1.8	1500	10000	2700	10000	4.7

SCHMITT TRIGGER CIRCUITS

SILICON 6 VOLT

VR	VT	RE	R2	RL	R1	LG
2.4	2.2	1500	10000	1000	12000	1.7
2.4	2.3	1500	10000	820	12000	1.4
2.6	1.3	1800	10000	10000	1500	18.8
2.6	1.4	1800	10000	8200	3300	15.4
2.6	1.5	1800	10000	6800	4700	12.8
2.6	1.6	1800	10000	5600	5600	10.8
2.6	1.7	1800	10000	4700	6800	8.8
2.6	1.9	1800	10000	3300	8200	6.2
2.6	2.3	1800	10000	1500	10000	2.8
2.6	2.4	1800	10000	1200	10000	2.3
2.8	1.3	1800	12000	10000	1800	19.2
2.8	1.4	1800	12000	8200	3300	16.0
2.8	1.5	1800	12000	6800	4700	13.3
2.8	1.8	1800	12000	4700	6800	9.2
2.8	2.0	1800	12000	3300	8200	6.5
2.8	2.3	1800	12000	1800	10000	3.5
2.8	2.4	1800	12000	1500	10000	2.9
3.0	1.5	2200	12000	10000	390	21.3
3.0	1.6	2200	12000	8200	2200	17.5
3.0	1.7	2200	12000	6800	3300	14.8
3.0	1.8	2200	12000	5600	4700	12.0
3.0	1.9	2200	12000	4700	5600	10.1
3.0	2.0	2200	12000	3900	6800	8.1
3.0	2.2	2200	12000	3300	6800	7.2
3.0	2.4	2200	12000	2200	8200	4.7
3.0	2.9	2200	12000	680	10000	1.4
3.2	1.5	2200	15000	10000	1200	20.9
3.2	1.6	2200	15000	8200	2700	17.5
3.2	1.7	2200	15000	6800	4700	13.9
3.2	1.9	2200	15000	5600	5600	11.7
3.2	2.0	2200	15000	4700	6800	9.6
3.2	2.2	2200	15000	3300	8200	6.8
3.2	2.4	2200	15000	2700	8200	5.8
3.2	2.7	2200	15000	1500	10000	3.1
3.2	2.8	2200	15000	1200	10000	2.5
3.2	3.1	2200	15000	820	10000	1.8
3.4	1.6	2200	15000	8200	1500	19.2
3.4	1.8	2200	15000	6800	2700	16.1
3.4	1.9	2200	15000	5600	3900	13.3
3.4	2.0	2200	15000	4700	4700	11.2

SCHMITT TRIGGER CIRCUITS
SILICON 6 VOLT

VR	VT	RE	R_2	RL	R_1	LG
3.4	2.2	2200	15000	3900	5600	9.3
3.4	2.5	2200	15000	2700	6800	6.4
3.4	2.7	2200	15000	1800	8200	4.1
3.4	2.8	2200	15000	1500	8200	3.5
3.4	3.0	2200	15000	1200	8200	2.9
3.6	1.7	2700	18000	10000	150	23.3
3.6	1.8	2700	18000	8200	1800	19.3
3.6	1.9	2700	18000	6800	3300	15.9
3.6	2.1	2700	18000	5600	4700	12.9
3.6	2.2	2700	18000	4700	5600	10.8
3.6	2.5	2700	18000	3300	6800	7.7
3.6	2.8	2700	18000	2200	8200	5.0
3.6	3.0	2700	18000	1800	8200	4.2
3.8	1.8	2700	18000	8200	560	21.3
3.8	2.0	2700	18000	6800	1800	17.9
3.8	2.1	2700	18000	5600	3300	14.4
3.8	2.3	2700	18000	4700	3900	12.4
3.8	2.5	2700	18000	3900	4700	10.3
3.8	2.6	2700	18000	3300	5600	8.5
3.8	2.9	2700	18000	2200	6800	5.6
3.8	3.1	2700	18000	1800	6800	4.7
3.8	3.5	2700	18000	820	8200	2.1
3.8	3.6	2700	18000	680	8200	1.7
3.8	3.7	2700	18000	560	8200	1.5
4.0	1.9	2700	22000	8200	820	21.4
4.0	2.0	2700	22000	6800	2200	17.8
4.0	2.2	2700	22000	5600	3300	14.7
4.0	2.7	2700	22000	3300	5600	8.7
4.0	3.0	2700	22000	2200	6800	5.7
4.0	3.6	2700	22000	1000	8200	2.6
4.0	3.7	2700	22000	820	8200	2.1
4.0	3.8	2700	22000	680	8200	1.8
4.0	3.9	2700	22000	560	8200	1.5
4.2	2.2	3300	22000	6800	1200	19.4
4.2	2.4	3300	22000	5600	2200	16.2
4.2	2.6	3300	22000	4700	3300	13.4
4.2	2.8	3300	22000	3900	3900	11.3
4.2	2.9	3300	22000	3300	4700	9.4
4.2	3.3	3300	22000	2200	5600	6.4
4.2	3.7	3300	22000	1200	6800	3.4

SCHMITT TRIGGER CIRCUITS

SILICON 6 VOLT

VR	VT	RE	R2	RL	R1	LG
4.2	3.8	3300	22000	1000	6800	2.9
4.2	3.9	3300	22000	820	6800	2.4
4.4	2.1	3300	27000	8200	100	23.2
4.4	2.3	3300	27000	6800	1200	19.8
4.4	2.5	3300	27000	5600	2200	16.6
4.4	2.6	3300	27000	4700	3300	13.7
4.4	2.8	3300	27000	3900	3900	11.5
4.4	3.0	3300	27000	3300	4700	9.6
4.4	3.4	3300	27000	2200	5600	6.5
4.4	3.7	3300	27000	1500	6800	4.3
4.4	3.8	3300	27000	1200	6800	3.5
4.4	4.0	3300	27000	1000	6800	3.0
4.4	4.1	3300	27000	820	6800	2.5
4.6	2.3	3300	33000	6800	1000	20.5
4.6	2.5	3300	33000	5600	2200	16.9
4.6	2.7	3300	33000	4700	3300	13.9
4.6	2.9	3300	33000	3900	3900	11.7
4.6	3.1	3300	33000	3300	4700	9.8
4.6	3.5	3300	33000	2200	5600	6.6
4.6	4.0	3300	33000	1200	6800	3.5
4.6	4.1	3300	33000	1000	6800	3.0
4.6	4.2	3300	33000	820	6800	2.5
4.6	4.4	3300	33000	680	6800	2.1
4.8	2.5	3900	39000	6800	820	21.1
4.8	2.7	3900	39000	5600	1800	17.7
4.8	2.9	3900	39000	4700	2700	14.9
4.8	3.1	3900	39000	3900	3900	11.9
4.8	3.3	3900	39000	3300	3900	10.6
4.8	3.5	3900	39000	2700	4700	8.5
4.8	3.7	3900	39000	2200	5600	6.7
4.8	3.9	3900	39000	1800	5600	5.7
4.8	4.3	3900	39000	1000	6800	3.1
4.8	4.4	3900	39000	820	6800	2.5
4.8	4.5	3900	39000	680	6800	2.1
4.8	4.7	3900	39000	470	6800	1.5
5.0	2.6	3900	47000	6800	150	22.6
5.0	2.8	3900	47000	5600	1200	18.9
5.0	3.0	3900	47000	4700	2200	15.7
5.0	3.2	3900	47000	3900	2700	13.4
5.0	3.4	3900	47000	3300	3300	11.4

SCHMITT TRIGGER CIRCUITS
SILICON 9 VOLT

VR	VT	RE	R2	RL	R1	LG
1.0	0.8	270	3300	10000	12000	6.3
1.0	0.9	270	3300	3900	18000	2.5
1.2	0.8	470	4700	22000	3900	14.9
1.2	0.9	470	4700	8200	18000	5.5
1.2	1.0	470	4700	3900	22000	2.6
1.2	1.1	470	4700	3300	22000	2.3
1.4	0.8	680	5600	22000	3900	16.5
1.4	0.9	680	5600	18000	8200	13.3
1.4	1.0	680	5600	8200	18000	6.1
1.4	1.1	680	5600	4700	22000	3.4
1.4	1.2	680	5600	3900	22000	2.9
1.6	0.9	820	5600	22000	470	19.0
1.6	1.0	820	5600	12000	10000	10.5
1.6	1.1	820	5600	8200	15000	6.9
1.6	1.2	820	5600	4700	18000	4.0
1.6	1.3	820	5600	3900	18000	3.4
1.8	0.9	1000	6800	22000	1500	19.7
1.8	1.0	1000	6800	18000	5600	16.1
1.8	1.1	1000	6800	12000	12000	10.5
1.8	1.2	1000	6800	8200	15000	7.4
1.8	1.3	1000	6800	5600	18000	5.0
1.8	1.6	1000	6800	2200	22000	1.9
1.8	1.7	1000	6800	1800	22000	1.6
2.0	1.0	1200	6800	18000	2700	18.2
2.0	1.1	1200	6800	15000	5600	15.2
2.0	1.5	1200	6800	5600	15000	5.7
2.0	1.6	1200	6800	3300	18000	3.3
2.0	1.7	1200	6800	2700	18000	2.7
2.2	1.0	1200	8200	22000	100	22.3
2.2	1.1	1200	8200	15000	6800	15.4
2.2	1.2	1200	8200	12000	10000	12.2
2.2	1.3	1200	8200	10000	12000	10.2
2.2	1.4	1200	8200	6800	15000	7.0
2.2	1.7	1200	8200	3900	18000	4.0
2.2	1.8	1200	8200	3300	18000	3.5
2.4	1.2	1500	8200	15000	4700	16.9
2.4	1.3	1500	8200	12000	8200	13.3
2.4	1.4	1500	8200	10000	10000	11.1
2.4	1.5	1500	8200	8200	12000	9.1
2.4	1.8	1500	8200	4700	15000	5.3

SCHMITT TRIGGER CIRCUITS

SILICON 9 VOLT

VR	VT	RE	R2	RL	R1	LG
2.4	2.1	1500	8200	2200	18000	2.4
2.4	2.2	1500	8200	1800	18000	2.0
2.4	2.3	1500	8200	1500	18000	1.7
2.6	1.2	1800	8200	18000	100	22.0
2.6	1.3	1800	8200	15000	3300	18.2
2.6	1.4	1800	8200	12000	5600	15.0
2.6	1.5	1800	8200	10000	8200	12.2
2.6	1.6	1800	8200	8200	10000	10.0
2.6	1.8	1800	8200	5600	12000	7.0
2.6	2.1	1800	8200	3300	15000	4.0
2.6	2.2	1800	8200	2700	15000	3.4
2.8	1.3	1800	8200	15000	1200	20.2
2.8	1.5	1800	8200	12000	3900	16.4
2.8	1.7	1800	8200	8200	8200	10.9
2.8	2.1	1800	8200	3900	12000	5.3
2.8	2.7	1800	8200	1200	15000	1.6
3.0	1.4	2200	10000	18000	100	23.3
3.0	1.5	2200	10000	15000	2700	19.8
3.0	1.6	2200	10000	12000	5600	15.9
3.0	1.7	2200	10000	10000	8200	12.9
3.0	1.8	2200	10000	8200	10000	10.6
3.0	2.1	2200	10000	5600	12000	7.4
3.0	2.4	2200	10000	3300	15000	4.2
3.0	2.5	2200	10000	2700	15000	3.6
3.2	1.5	2200	10000	15000	1200	21.4
3.2	1.6	2200	10000	12000	3900	17.4
3.2	1.9	2200	10000	8200	8200	11.6
3.2	2.4	2200	10000	3900	12000	5.7
3.2	2.9	2200	10000	1500	15000	2.1
3.2	3.0	2200	10000	1200	15000	1.7
3.2	3.1	2200	10000	1000	15000	1.4
3.4	1.7	2200	10000	12000	2700	18.6
3.4	1.8	2200	10000	10000	4700	15.5
3.4	1.9	2200	10000	8200	6800	12.5
3.4	2.0	2200	10000	6800	8200	10.3
3.4	2.3	2200	10000	4700	10000	7.3
3.4	2.7	2200	10000	2700	12000	4.2
3.6	1.8	2700	10000	12000	1500	20.0
3.6	2.0	2700	10000	10000	3300	16.9
3.6	2.1	2700	10000	8200	5600	13.4

VR	VT	RE	R2	RL	R1	LG
3.6	2.3	2700	10000	6800	6800	11.3
3.6	2.4	2700	10000	5600	8200	9.2
3.6	2.8	2700	10000	3300	10000	5.6
3.6	3.2	2700	10000	1800	12000	2.9
3.6	3.3	2700	10000	1500	12000	2.5
3.8	1.9	2700	12000	12000	2700	19.4
3.8	2.0	2700	12000	10000	4700	16.2
3.8	2.6	2700	12000	4700	10000	7.6
3.8	3.1	2700	12000	2700	12000	4.4
4.0	1.9	2700	12000	12000	1500	20.8
4.0	2.0	2700	12000	10000	3300	17.6
4.0	2.4	2700	12000	6800	6800	11.7
4.0	3.0	2700	12000	3300	10000	5.8
4.0	3.6	2700	12000	1500	12000	2.6
4.0	3.7	2700	12000	1200	12000	2.1
4.2	2.1	3300	12000	12000	390	22.4
4.2	2.3	3300	12000	10000	2200	18.9
4.2	2.6	3300	12000	6800	5600	12.7
4.2	2.8	3300	12000	5600	6800	10.5
4.2	3.6	3300	12000	2200	10000	4.2
4.2	4.1	3300	12000	680	12000	1.2
4.4	2.3	3300	12000	10000	1200	20.2
4.4	2.5	3300	12000	8200	3300	16.3
4.4	2.7	3300	12000	6800	4700	13.5
4.4	2.9	3300	12000	5600	5600	11.3
4.4	3.0	3300	12000	4700	6800	9.3
4.4	3.3	3300	12000	3300	8200	6.5
4.4	3.9	3300	12000	1500	10000	3.0
4.4	4.1	3300	12000	1200	10000	2.4
4.6	2.2	3300	15000	12000	680	23.0
4.6	2.3	3300	15000	10000	2700	19.1
4.6	2.5	3300	15000	8200	4700	15.5
4.6	3.1	3300	15000	4700	8200	8.9
4.6	3.7	3300	15000	2700	10000	5.2
4.6	4.3	3300	15000	1000	12000	1.9
4.6	4.4	3300	15000	820	12000	1.6
4.6	4.5	3300	15000	680	12000	1.3
4.8	2.5	3900	15000	10000	1800	20.3
4.8	4.2	3900	15000	1800	10000	3.7
5.0	2.6	3900	15000	10000	820	21.7

SCHMITT TRIGGER CIRCUITS

SILICON 9 VOLT

VR	VT	RE	R2	RL	R1	LG
5.0	2.8	3900	15000	8200	2700	17.7
5.0	3.0	3900	15000	6800	3900	14.9
5.0	3.6	3900	15000	3900	6800	8.5
5.0	4.0	3900	15000	2700	8200	5.8
5.0	4.7	3900	15000	1000	10000	2.1
5.0	4.8	3900	15000	820	10000	1.8
5.0	4.9	3900	15000	680	10000	1.5
5.2	2.6	3900	18000	10000	1800	20.9
5.2	2.9	3900	18000	8200	3300	17.5
5.2	3.1	3900	18000	6800	4700	14.5
5.2	3.5	3900	18000	4700	6800	10.0
5.2	3.9	3900	18000	3300	8200	7.0
5.2	4.5	3900	18000	1800	10000	3.8
5.2	4.7	3900	18000	1500	10000	3.2
5.4	2.7	3900	18000	10000	560	22.8
5.4	2.9	3900	18000	8200	2200	18.9
5.4	3.1	3900	18000	6800	3900	15.3
5.4	3.8	3900	18000	3900	6800	8.8
5.4	4.5	3900	18000	2200	8200	5.1
5.4	5.1	3900	18000	820	10000	1.8
5.4	5.2	3900	18000	680	10000	1.5
5.4	5.3	3900	18000	560	10000	1.3
5.6	3.2	4700	18000	8200	1500	19.9
5.6	4.9	4700	18000	1800	8200	4.3
5.6	5.0	4700	18000	1500	8200	3.6
5.8	3.2	4700	18000	8200	680	21.3
5.8	3.5	4700	18000	6800	2200	17.5
5.8	3.7	4700	18000	5600	3300	14.5
5.8	4.4	4700	18000	3300	5600	8.5
5.8	4.8	4700	18000	2200	6800	5.6
5.8	5.5	4700	18000	820	8200	2.1
5.8	5.6	4700	18000	680	8200	1.8
5.8	5.7	4700	18000	560	8200	1.5
6.0	3.3	4700	22000	8200	1500	20.4
6.0	3.6	4700	22000	6800	2700	17.2
6.0	3.8	4700	22000	5600	3900	14.2
6.0	4.3	4700	22000	3900	5600	9.9
6.0	4.8	4700	22000	2700	6800	6.8
6.0	5.3	4700	22000	1500	8200	3.7
6.2	3.3	4700	22000	8200	560	22.0

VR	VT	RE	R2	RL	R1	LG
1.00	0.75	270	3300	27000	3300	12.7
1.25	0.75	470	4700	33000	1200	17.3
1.25	1.00	470	4700	6800	27000	3.6
1.50	0.75	680	5600	33000	470	19.5
1.50	1.00	680	5600	12000	22000	7.0
1.50	1.25	680	5600	5600	27000	3.4
1.75	1.00	1000	5600	22000	6800	15.3
1.75	1.25	1000	5600	8200	22000	5.4
1.75	1.50	1000	5600	3300	27000	2.2
2.00	1.00	1200	6800	27000	2700	19.7
2.00	1.25	1200	6800	12000	18000	8.7
2.00	1.50	1200	6800	6800	22000	5.1
2.00	1.75	1200	6800	3300	27000	2.4
2.25	1.00	1500	6800	27000	100	21.6
2.25	1.25	1500	6800	18000	8200	14.8
2.25	1.50	1500	6800	8200	18000	6.8
2.25	1.75	1500	6800	4700	22000	3.8
2.25	2.00	1500	6800	3300	22000	2.8
2.50	1.25	1500	6800	18000	5600	16.3
2.50	1.50	1500	6800	12000	10000	11.6
2.50	1.75	1500	6800	8200	15000	7.6
2.50	2.00	1500	6800	4700	18000	4.4
2.50	2.25	1500	6800	1800	22000	1.6
2.75	1.25	1800	8200	22000	2700	20.4
2.75	1.50	1800	8200	15000	10000	13.8
2.75	1.75	1800	8200	10000	15000	9.2
2.75	2.00	1800	8200	6800	18000	6.3
2.75	2.25	1800	8200	3300	22000	3.0
2.75	2.50	1800	8200	2700	22000	2.5
3.00	1.50	2200	8200	18000	4700	18.1
3.00	1.75	2200	8200	12000	10000	12.4
3.00	2.00	2200	8200	8200	15000	8.1
3.00	2.25	2200	8200	4700	18000	4.7
3.00	2.50	2200	8200	3900	18000	4.1
3.00	2.75	2200	8200	1500	22000	1.5
3.25	1.50	2200	8200	18000	2200	20.1
3.25	1.75	2200	8200	12000	8200	13.4
3.25	2.00	2200	8200	10000	10000	11.3
3.25	2.25	2200	8200	5600	15000	6.1
3.25	2.50	2200	8200	4700	15000	5.4

SCHMITT TRIGGER CIRCUITS

SILICON 12 VOLT

VR	VT	RE	R2	RL	R1	LG
3.25	2.75	2200	8200	2700	18000	3.0
3.25	3.00	2200	8200	1800	18000	2.0
3.50	1.75	2700	8200	18000	330	22.0
3.50	2.00	2700	8200	12000	6800	14.3
3.50	2.25	2700	8200	8200	10000	10.1
3.50	2.50	2700	8200	6800	12000	8.1
3.50	2.75	2700	8200	3900	15000	4.6
3.50	3.00	2700	8200	3300	15000	4.0
3.50	3.25	2700	8200	1200	18000	1.4
3.75	1.75	2700	10000	18000	2200	21.3
3.75	2.00	2700	10000	12000	8200	14.2
3.75	2.25	2700	10000	10000	10000	11.9
3.75	2.50	2700	10000	8200	12000	9.7
3.75	2.75	2700	10000	5600	15000	6.5
3.75	3.25	2700	10000	2200	18000	2.6
3.75	3.50	2700	10000	1800	18000	2.2
4.00	1.75	2700	10000	18000	180	23.4
4.00	2.00	2700	10000	15000	3300	19.4
4.00	2.25	2700	10000	10000	8200	13.0
4.00	2.50	2700	10000	8200	10000	10.6
4.00	2.75	2700	10000	5600	12000	7.5
4.00	3.00	2700	10000	3900	15000	4.9
4.00	3.25	2700	10000	3300	15000	4.3
4.00	3.50	2700	10000	2700	15000	3.6
4.25	2.25	3300	10000	15000	1800	20.9
4.25	2.50	3300	10000	10000	6800	13.9
4.25	2.75	3300	10000	8200	8200	11.7
4.25	3.00	3300	10000	6800	10000	9.5
4.25	3.25	3300	10000	4700	12000	6.6
4.25	3.75	3300	10000	2200	15000	3.0
4.25	4.00	3300	10000	1500	15000	2.1
4.50	2.25	3300	10000	15000	330	22.6
4.50	2.50	3300	10000	12000	3300	18.1
4.50	2.75	3300	10000	8200	6800	12.6
4.50	3.00	3300	10000	6800	8200	10.4
4.50	3.25	3300	10000	5600	10000	8.3
4.50	3.50	3300	10000	3900	12000	5.7
4.50	3.75	3300	10000	3300	12000	5.0
4.50	4.00	3300	10000	2700	12000	4.2
4.50	4.25	3300	10000	1000	15000	1.5

SCHMITT TRIGGER CIRCUITS
SILICON 12 VOLT

VR	VT	RE	R2	RL	R1	LG
4.75	2.50	3900	12000	15000	1800	21.9
4.75	2.75	3900	12000	10000	6800	14.6
4.75	3.00	3900	12000	8200	8200	12.2
4.75	3.25	3900	12000	6800	10000	9.9
4.75	3.75	3900	12000	4700	12000	6.9
4.75	4.25	3900	12000	2200	15000	3.1
4.75	4.50	3900	12000	1500	15000	2.2
5.00	2.50	3900	12000	15000	470	23.4
5.00	2.75	3900	12000	12000	3300	18.9
5.00	3.00	3900	12000	10000	5600	15.5
5.00	3.25	3900	12000	8200	6800	13.1
5.00	3.50	3900	12000	5600	10000	8.7
5.00	4.00	3900	12000	3300	12000	5.2
5.00	4.75	3900	12000	1000	15000	1.5
5.25	2.75	3900	12000	12000	2200	20.1
5.25	3.00	3900	12000	10000	3900	17.1
5.25	3.25	3900	12000	8200	5600	14.1
5.25	3.50	3900	12000	6800	6800	11.8
5.25	3.75	3900	12000	5600	8200	9.6
5.25	4.00	3900	12000	3900	10000	6.7
5.25	4.25	3900	12000	2700	12000	4.4
5.25	4.50	3900	12000	2200	12000	3.7
5.25	4.75	3900	12000	1800	12000	3.1
5.50	3.00	4700	12000	12000	1200	21.4
5.50	3.25	4700	12000	10000	3300	17.7
5.50	3.50	4700	12000	8200	4700	14.9
5.50	3.75	4700	12000	6800	6800	11.8
5.50	4.25	4700	12000	4700	8200	8.5
5.50	4.50	4700	12000	3300	10000	5.9
5.50	4.75	4700	12000	2700	10000	5.0
5.50	5.00	4700	12000	1500	12000	2.6
5.50	5.25	4700	12000	1200	12000	2.1
5.75	3.00	4700	12000	12000	120	22.9
5.75	3.25	4700	12000	10000	2200	19.0
5.75	3.50	4700	12000	8200	3900	15.7
5.75	3.75	4700	12000	6800	5600	12.8
5.75	4.00	4700	12000	5600	6800	10.5
5.75	4.50	4700	12000	3900	8200	7.5
5.75	5.00	4700	12000	2200	10000	4.2
5.75	5.25	4700	12000	1800	10000	3.5

SCHMITT TRIGGER CIRCUITS

SILICON 12 VOLT

VR	VT	RE	R2	RL	R1	LG
6.00	3.25	4700	15000	12000	1800	21.6
6.00	3.50	4700	15000	10000	3900	17.9
6.00	3.75	4700	15000	8200	5600	14.7
6.00	4.00	4700	15000	6800	6800	12.4
6.00	4.25	4700	15000	5600	8200	10.1
6.00	4.75	4700	15000	3900	10000	7.0
6.00	5.00	4700	15000	3300	10000	6.1
6.00	5.50	4700	15000	1800	12000	3.2
6.00	5.75	4700	15000	1200	12000	2.2
6.25	3.25	4700	15000	12000	680	23.1
6.25	3.50	4700	15000	10000	2700	19.2
6.25	3.75	4700	15000	8200	4700	15.6
6.25	4.00	4700	15000	6800	5600	13.3
6.25	4.25	4700	15000	5600	6800	11.0
6.25	4.50	4700	15000	4700	8200	8.9
6.25	5.25	4700	15000	2700	10000	5.2
6.25	5.50	4700	15000	2200	10000	4.4
6.25	6.00	4700	15000	820	12000	1.6
6.50	3.50	5600	15000	12000	100	24.0
6.50	3.75	5600	15000	10000	1800	20.4
6.50	4.00	5600	15000	8200	3900	16.4
6.50	4.50	5600	15000	6800	4700	14.1
6.50	5.00	5600	15000	4700	6800	9.8
6.50	5.25	5600	15000	3300	8200	6.9
6.50	5.75	5600	15000	1800	10000	3.7
6.50	6.00	5600	15000	1500	10000	3.1
6.75	3.75	5600	15000	10000	820	21.8
6.75	4.25	5600	15000	8200	2700	17.8
6.75	4.50	5600	15000	6800	3900	15.0
6.75	4.75	5600	15000	5600	5600	11.9
6.75	5.25	5600	15000	3900	6800	8.6
6.75	5.75	5600	15000	2700	8200	5.9
6.75	6.00	5600	15000	2200	8200	4.9
6.75	6.25	5600	15000	1200	10000	2.5
6.75	6.50	5600	15000	820	10000	1.8
7.00	3.50	5600	18000	12000	100	24.7
7.00	4.00	5600	18000	10000	1800	21.0
7.00	4.25	5600	18000	8200	3300	17.6
7.00	4.50	5600	18000	6800	4700	14.6
7.00	5.25	5600	18000	4700	6800	10.1

SCHMITT TRIGGER CIRCUITS
SILICON 24 VOLT

VR	VT	RE	R2	RL	R1	LG
1.5	1.0	680	4700	22000	39000	6.8
2.0	1.0	1200	5600	56000	100	20.9
2.0	1.5	1200	5600	15000	39000	5.8
2.5	1.0	1500	6800	47000	5600	20.2
2.5	1.5	1500	6800	22000	33000	9.1
2.5	2.0	1500	6800	6800	47000	2.9
3.0	1.5	2200	6800	39000	4700	20.2
3.0	2.0	2200	6800	18000	27000	9.1
3.0	2.5	2200	6800	6800	39000	3.4
3.5	1.5	2700	6800	39000	100	22.5
3.5	2.0	2700	6800	27000	10000	16.4
3.5	2.5	2700	6800	15000	22000	9.1
3.5	3.0	2700	6800	5600	33000	3.3
4.0	1.5	2700	8200	39000	100	24.0
4.0	2.0	2700	8200	27000	12000	16.6
4.0	2.5	2700	8200	15000	22000	9.7
4.0	3.0	2700	8200	10000	27000	6.5
4.0	3.5	2700	8200	4700	33000	3.0
4.5	2.0	3300	8200	33000	220	23.7
4.5	2.5	3300	8200	22000	12000	15.5
4.5	3.0	3300	8200	15000	18000	10.8
4.5	3.5	3300	8200	6800	27000	4.8
4.5	4.0	3300	8200	4700	27000	3.5
5.0	2.5	3900	8200	27000	2200	21.9
5.0	3.0	3900	8200	18000	12000	14.2
5.0	3.5	3900	8200	12000	18000	9.5
5.0	4.0	3900	8200	6800	22000	5.6
5.0	4.5	3900	8200	3300	27000	2.6
5.5	3.0	4700	8200	22000	4700	19.4
5.5	3.5	4700	8200	18000	8200	16.1
5.5	4.0	4700	8200	12000	15000	10.5
5.5	4.5	4700	8200	8200	18000	7.4
5.5	5.0	4700	8200	3900	22000	3.5
6.0	3.5	4700	8200	18000	5600	17.7
6.0	4.0	4700	8200	12000	12000	11.6
6.0	4.5	4700	8200	8200	15000	8.2
6.0	5.0	4700	8200	5600	18000	5.5
6.0	5.5	4700	8200	2200	22000	2.1
6.5	3.5	5600	10000	22000	3900	21.1
6.5	4.0	5600	10000	18000	6800	18.0

SCHMITT TRIGGER CIRCUITS
SILICON 24 VOLT

VR	VT	RE	R2	RL	R1	LG
6.5	5.0	5600	10000	8200	18000	7.8
6.5	5.5	5600	10000	4700	22000	4.4
6.5	6.0	5600	10000	3300	22000	3.2
7.0	4.0	5600	10000	18000	5600	18.8
7.0	4.5	5600	10000	15000	8200	15.9
7.0	5.5	5600	10000	8200	15000	8.7
7.0	6.0	5600	10000	4700	18000	5.1
7.0	6.5	5600	10000	1800	22000	1.9
7.5	4.5	5600	10000	15000	5600	17.6
7.5	5.0	5600	10000	12000	8200	14.4
7.5	5.5	5600	10000	8200	12000	9.8
7.5	6.0	5600	10000	5600	15000	6.6
7.5	6.5	5600	10000	3300	18000	3.8
7.5	7.0	5600	10000	2700	18000	3.2
8.0	4.5	6800	10000	18000	1200	22.6
8.0	5.0	6800	10000	15000	3900	19.1
8.0	5.5	6800	10000	12000	6800	15.3
8.0	6.5	6800	10000	6800	12000	8.7
8.0	7.0	6800	10000	3900	15000	5.0
8.0	7.5	6800	10000	1800	18000	2.2
8.5	5.0	6800	10000	15000	2700	20.2
8.5	5.5	6800	10000	12000	5600	16.2
8.5	6.0	6800	10000	10000	8200	13.1
8.5	7.0	6800	10000	5600	12000	7.6
8.5	7.5	6800	10000	2700	15000	3.6
8.5	8.0	6800	10000	2200	15000	3.0
9.0	5.0	6800	12000	18000	1000	23.8
9.0	5.5	6800	12000	15000	3900	19.9
9.0	6.0	6800	12000	12000	6800	16.0
9.0	7.0	6800	12000	6800	12000	9.1
9.0	7.5	6800	12000	4700	15000	6.0
9.0	8.0	6800	12000	3900	15000	5.2
9.0	8.5	6800	12000	1500	18000	1.9
9.5	5.5	8200	12000	18000	100	24.8
9.5	6.0	8200	12000	15000	2700	21.1
9.5	6.5	8200	12000	12000	5600	16.9
9.5	7.0	8200	12000	10000	6800	14.7
9.5	8.0	8200	12000	5600	12000	7.9
9.5	8.5	8200	12000	2700	15000	3.8
9.5	9.0	8200	12000	2200	15000	3.2

SCHMITT TRIGGER CIRCUITS
SILICON 24 VOLT

VR	VT	RE	R2	RL	R1	LG
10.0	6.0	8200	12000	15000	1200	22.7
10.0	6.5	8200	12000	12000	3900	18.5
10.0	7.5	8200	12000	8200	8200	12.3
10.0	9.0	8200	12000	3900	12000	6.0
10.0	9.5	8200	12000	1500	15000	2.2
10.5	6.5	8200	12000	15000	100	24.1
10.5	7.0	8200	12000	12000	2700	19.7
10.5	7.5	8200	12000	10000	4700	16.4
10.5	8.0	8200	12000	6800	8200	11.0
10.5	9.0	8200	12000	4700	10000	7.7
10.5	9.5	8200	12000	2700	12000	4.4
10.5	10.0	8200	12000	2200	12000	3.7
11.0	7.5	10000	12000	12000	1800	20.8
11.0	8.0	10000	12000	10000	3900	17.2
11.0	8.5	10000	12000	8200	5600	14.2
11.0	9.0	10000	12000	5600	8200	9.7
11.0	9.5	10000	12000	3900	10000	6.7
11.0	10.0	10000	12000	3300	10000	5.9
11.0	10.5	10000	12000	1800	12000	3.1
11.5	7.0	10000	15000	15000	680	24.4
11.5	8.0	10000	15000	10000	5600	16.4
11.5	9.0	10000	15000	8200	6800	13.8
11.5	9.5	10000	15000	5600	10000	9.2
11.5	10.0	10000	15000	3900	12000	6.3
11.5	10.5	10000	15000	3300	12000	5.5
11.5	11.0	10000	15000	1200	15000	1.9
12.0	8.0	10000	15000	12000	2200	21.2
12.0	8.5	10000	15000	10000	4700	17.2
12.0	9.0	10000	15000	8200	5600	14.8
12.0	10.0	10000	15000	4700	10000	8.1
12.0	10.5	10000	15000	3900	10000	7.0
12.0	11.0	10000	15000	2200	12000	3.9
12.0	11.5	10000	15000	1800	12000	3.3
12.5	8.0	10000	15000	12000	1200	22.5
12.5	8.5	10000	15000	10000	3300	18.6
12.5	9.5	10000	15000	6800	6800	12.4
12.5	10.5	10000	15000	4700	8200	9.0
12.5	11.0	10000	15000	3300	10000	6.1
12.5	11.5	10000	15000	2700	10000	5.2
12.5	12.0	10000	15000	1200	12000	2.2

SCHMITT TRIGGER CIRCUITS
SILICON 24 VOLT

VR	VT	RE	R2	RL	R1	LG
13.0	9.0	12000	15000	12000	220	23.9
13.0	9.5	12000	15000	10000	2200	20.0
13.0	10.0	12000	15000	8200	3900	16.5
13.0	10.5	12000	15000	6800	5600	13.4
13.0	11.5	12000	15000	3900	8200	7.8
13.0	12.0	12000	15000	2200	10000	4.4
13.0	12.5	12000	15000	1800	10000	3.7
13.5	9.0	12000	18000	12000	1500	22.8
13.5	9.5	12000	18000	10000	3300	19.2
13.5	10.0	12000	18000	8200	5600	15.3
13.5	10.5	12000	18000	6800	6800	12.8
13.5	11.0	12000	18000	5600	8200	10.4
13.5	11.5	12000	18000	4700	8200	9.2
13.5	12.0	12000	18000	3300	10000	6.3
13.5	12.5	12000	18000	1800	12000	3.4
13.5	13.0	12000	18000	1500	12000	2.8
14.0	9.5	12000	18000	12000	270	24.6
14.0	10.0	12000	18000	10000	2200	20.6
14.0	10.5	12000	18000	8200	3900	17.0
14.0	11.0	12000	18000	6800	5600	13.8
14.0	11.5	12000	18000	5600	6800	11.4
14.0	12.0	12000	18000	3900	8200	8.1
14.0	12.5	12000	18000	2700	10000	5.4
14.0	13.0	12000	18000	2200	10000	4.5
14.0	13.5	12000	18000	1800	10000	3.8
14.5	10.0	12000	18000	10000	1200	22.0
14.5	10.5	12000	18000	8200	3300	17.6
14.5	11.0	12000	18000	6800	4700	14.6
14.5	12.0	12000	18000	4700	6800	10.1
14.5	12.5	12000	18000	3300	8200	7.1
14.5	14.0	12000	18000	1200	10000	2.6
15.0	10.5	12000	18000	10000	330	23.3
15.0	11.0	12000	18000	8200	2200	19.0
15.0	11.5	12000	18000	6800	3300	16.1
15.0	12.0	12000	18000	5600	4700	13.1
15.0	12.5	12000	18000	4700	5600	11.0
15.0	14.0	12000	18000	2200	8200	5.1
15.5	10.5	12000	22000	10000	1500	22.1
15.5	11.0	12000	22000	8200	3300	18.1
15.5	12.0	12000	22000	6800	4700	15.0

VR	VT	RE	R2	RL	R1	LG
1.0	0.8	680	6800	2700	22000	1.7
1.0	0.9	680	6800	2200	22000	1.4
1.2	0.8	1000	8200	4700	22000	3.0
1.2	0.9	1000	8200	3300	22000	2.2
1.2	1.0	1000	8200	2700	22000	1.8
1.2	1.1	1000	8200	2200	22000	1.5
1.4	0.8	1200	8200	6800	15000	5.2
1.4	1.0	1200	8200	3900	18000	3.0
1.6	0.8	1200	8200	6800	12000	6.0
1.6	0.9	1200	8200	5600	12000	5.2
1.6	1.1	1200	8200	3300	15000	3.0
1.6	1.2	1200	8200	2700	15000	2.5
1.8	0.8	1500	8200	10000	5600	10.4
1.8	1.0	1500	8200	5600	10000	5.8
1.8	1.2	1500	8200	3900	12000	4.0
1.8	1.6	1500	8200	1500	15000	1.5
1.8	1.7	1500	8200	1000	15000	1.0
2.0	0.8	1500	8200	10000	3900	11.5
2.0	0.9	1500	8200	8200	5600	9.5
2.0	1.0	1500	8200	6800	6800	7.9
2.0	1.1	1500	8200	5600	8200	6.5
2.0	1.2	1500	8200	3900	10000	4.5
2.0	1.4	1500	8200	3300	10000	3.9
2.0	1.6	1500	8200	1800	12000	2.1
2.0	1.7	1500	8200	1500	12000	1.8
2.2	0.8	1800	10000	12000	2200	14.3
2.2	0.9	1800	10000	10000	4700	11.6
2.2	1.3	1800	10000	4700	10000	5.5
2.2	1.4	1800	10000	3900	10000	4.7
2.2	1.6	1800	10000	2700	12000	3.1
2.2	1.7	1800	10000	2200	12000	2.6
2.4	0.8	1800	10000	12000	470	16.0
2.4	0.9	1800	10000	10000	2700	13.1
2.4	1.0	1800	10000	8200	4700	10.6
2.4	1.1	1800	10000	6800	5600	9.1
2.4	1.3	1800	10000	5600	6800	7.5
2.4	1.5	1800	10000	3900	8200	5.3
2.4	1.7	1800	10000	2700	10000	3.5
2.4	1.8	1800	10000	2200	10000	3.0
2.6	1.0	2200	10000	10000	1200	14.6

SCHMITT TRIGGER CIRCUITS
GERMANIUM 6 VOLT

VR	VT	RE	R2	RL	R1	LG
2.6	1.2	2200	10000	8200	2700	12.2
2.6	1.4	2200	10000	5600	5600	8.2
2.6	1.5	2200	10000	4700	6800	6.7
2.6	1.8	2200	10000	3300	8200	4.7
2.6	1.9	2200	10000	2700	8200	4.0
2.6	2.2	2200	10000	1500	10000	2.1
2.6	2.3	2200	10000	1200	10000	1.8
2.6	2.4	2200	10000	1000	10000	1.5
2.6	2.5	2200	10000	820	10000	1.2
2.8	1.1	2200	12000	10000	1500	14.9
2.8	1.2	2200	12000	8200	3300	12.2
2.8	1.3	2200	12000	6800	4700	10.1
2.8	1.5	2200	12000	5600	5600	8.5
2.8	1.6	2200	12000	4700	6800	7.0
2.8	1.8	2200	12000	3300	8200	4.9
2.8	2.2	2200	12000	1800	10000	2.6
2.8	2.4	2200	12000	1500	10000	2.2
2.8	2.5	2200	12000	1200	10000	1.8
3.0	1.2	2700	12000	10000	270	16.3
3.0	1.4	2700	12000	8200	2200	13.3
3.0	1.5	2700	12000	6800	3300	11.2
3.0	1.7	2700	12000	5600	4700	9.1
3.0	1.8	2700	12000	4700	5600	7.7
3.0	2.1	2700	12000	3300	6800	5.5
3.0	2.3	2700	12000	2200	8200	3.6
3.0	2.5	2700	12000	1800	8200	3.0
3.2	1.4	2700	12000	8200	820	14.8
3.2	1.5	2700	12000	6800	2200	12.3
3.2	1.7	2700	12000	5600	3300	10.2
3.2	2.1	2700	12000	3300	5600	6.0
3.2	2.4	2700	12000	2200	6800	4.0
3.2	2.9	2700	12000	1000	8200	1.8
3.2	3.0	2700	12000	820	8200	1.5
3.2	3.1	2700	12000	680	8200	1.2
3.4	1.4	2700	15000	8200	1200	14.9
3.4	1.6	2700	15000	6800	2700	12.3
3.4	1.8	2700	15000	5600	3900	10.1
3.4	1.9	2700	15000	4700	4700	8.5
3.4	2.1	2700	15000	3900	5600	7.0
3.4	2.4	2700	15000	2700	6800	4.9

VR	VT	RE	R2	RL	R1	LG
3.4	2.8	2700	15000	1500	8200	2.7
3.4	3.0	2700	15000	1200	8200	2.2
3.4	3.1	2700	15000	1000	8200	1.8
3.6	1.6	3300	15000	8200	330	16.1
3.6	1.8	3300	15000	6800	1800	13.2
3.6	2.0	3300	15000	5600	2700	11.2
3.6	2.1	3300	15000	4700	3900	9.2
3.6	2.3	3300	15000	3900	4700	7.6
3.6	2.6	3300	15000	2700	5600	5.4
3.6	2.9	3300	15000	1800	6800	3.5
3.6	3.1	3300	15000	1500	6800	3.0
3.6	3.5	3300	15000	560	8200	1.1
3.8	1.7	3300	18000	8200	470	16.3
3.8	1.8	3300	18000	6800	1800	13.6
3.8	2.0	3300	18000	5600	3300	10.9
3.8	2.2	3300	18000	4700	3900	9.4
3.8	2.4	3300	18000	3900	4700	7.8
3.8	2.5	3300	18000	3300	5600	6.4
3.8	3.1	3300	18000	1800	6800	3.6
3.8	3.6	3300	18000	680	8200	1.3
3.8	3.7	3300	18000	560	8200	1.1
4.0	1.9	3300	18000	6800	680	15.0
4.0	2.1	3300	18000	5600	1800	12.5
4.0	2.2	3300	18000	4700	2700	10.5
4.0	2.5	3300	18000	3900	3300	8.8
4.0	2.6	3300	18000	3300	3900	7.5
4.0	2.8	3300	18000	2700	4700	6.0
4.0	3.2	3300	18000	1800	5600	4.0
4.0	3.7	3300	18000	820	6800	1.8
4.0	3.8	3300	18000	680	6800	1.5
4.0	3.9	3300	18000	560	6800	1.2
4.2	2.1	3900	22000	6800	1000	15.0
4.2	2.3	3900	22000	5600	2200	12.3
4.2	2.7	3900	22000	3900	3900	8.6
4.2	2.8	3900	22000	3300	4700	7.1
4.2	3.2	3900	22000	2200	5600	4.8
4.2	3.7	3900	22000	1200	6800	2.6
4.2	3.8	3900	22000	1000	6800	2.2
4.2	3.9	3900	22000	820	6800	1.8
4.2	4.0	3900	22000	680	6800	1.5

SCHMITT TRIGGER CIRCUITS

GERMANIUM 6 VOLT

VR	VT	RE	R_2	RL	R_1	LG
4.4	2.3	3900	22000	5600	1000	13.8
4.4	2.5	3900	22000	4700	1800	11.7
4.4	2.7	3900	22000	3900	2700	9.6
4.4	2.9	3900	22000	3300	3300	8.1
4.4	3.1	3900	22000	2700	3900	6.7
4.4	3.5	3900	22000	1800	4700	4.5
4.4	3.8	3900	22000	1200	5600	2.9
4.4	3.9	3900	22000	1000	5600	2.5
4.4	4.1	3900	22000	820	5600	2.1
4.6	2.4	3900	27000	5600	820	14.3
4.6	2.6	3900	27000	4700	1800	11.9
4.6	2.8	3900	27000	3900	2700	9.8
4.6	3.0	3900	27000	3300	3300	8.3
4.6	3.2	3900	27000	2700	3900	6.8
4.6	3.6	3900	27000	1800	4700	4.6
4.6	3.8	3900	27000	1500	4700	3.9
4.6	3.9	3900	27000	1200	5600	3.0
4.6	4.1	3900	27000	1000	5600	2.5
4.6	4.2	3900	27000	820	5600	2.1
4.6	4.3	3900	27000	680	5600	1.8
4.8	2.4	3900	33000	5600	680	14.8
4.8	2.6	3900	33000	4700	1500	12.5
4.8	2.9	3900	33000	3900	2200	10.5
4.8	3.1	3900	33000	3300	2700	8.9
4.8	3.3	3900	33000	2700	3300	7.3
4.8	3.5	3900	33000	2200	3900	5.9
4.8	3.7	3900	33000	1800	4700	4.6
4.8	3.9	3900	33000	1500	4700	4.0
4.8	4.3	3900	33000	820	5600	2.1
4.8	4.5	3900	33000	680	5600	1.8
4.8	4.6	3900	33000	560	5600	1.5
4.8	4.7	3900	33000	470	5600	1.3
5.0	2.7	4700	39000	5600	270	15.6
5.0	2.9	4700	39000	4700	1200	13.1
5.0	3.1	4700	39000	3900	2200	10.6
5.0	3.2	4700	39000	3900	1800	11.1
5.0	3.4	4700	39000	3300	2700	9.1
5.0	3.6	4700	39000	2700	3300	7.4
5.0	3.8	4700	39000	2200	3900	6.0
5.0	4.0	4700	39000	1800	3900	5.1

SCHMITT TRIGGER CIRCUITS
GERMANIUM 9 VOLT

VR	VT	RE	R2	RL	R1	LG
1.0	0.9	680	6800	2700	39000	1.0
1.2	1.0	1000	6800	3300	33000	1.5
1.2	1.1	1000	6800	2700	33000	1.2
1.4	0.9	1200	6800	8200	22000	4.4
1.4	1.0	1200	6800	4700	27000	2.4
1.4	1.1	1200	6800	3900	27000	2.1
1.4	1.2	1200	6800	3300	27000	1.8
1.4	1.3	1200	6800	2700	27000	1.5
1.6	0.9	1200	6800	8200	18000	5.0
1.6	1.1	1200	6800	4700	22000	2.8
1.6	1.2	1200	6800	3900	22000	2.4
1.6	1.3	1200	6800	3300	22000	2.1
1.8	0.8	1500	6800	15000	8200	10.4
1.8	1.0	1500	6800	8200	15000	5.7
1.8	1.2	1500	6800	5600	18000	3.8
1.8	1.3	1500	6800	4700	18000	3.3
1.8	1.7	1500	6800	1500	22000	1.0
2.0	0.9	1500	8200	12000	12000	8.6
2.0	1.3	1500	8200	5600	18000	4.1
2.0	1.6	1500	8200	2700	22000	1.9
2.0	1.7	1500	8200	2200	22000	1.6
2.0	1.8	1500	8200	1800	22000	1.3
2.0	1.9	1500	8200	1500	22000	1.1
2.2	0.8	1800	8200	18000	3900	14.0
2.2	0.9	1800	8200	15000	6800	11.7
2.2	1.0	1800	8200	12000	10000	9.3
2.2	1.1	1800	8200	10000	12000	7.8
2.2	1.3	1800	8200	6800	15000	5.3
2.2	1.6	1800	8200	3900	18000	3.0
2.2	1.8	1800	8200	3300	18000	2.6
2.4	0.8	1800	8200	18000	1500	15.6
2.4	0.9	1800	8200	15000	4700	12.8
2.4	1.6	1800	8200	4700	15000	4.0
2.4	1.8	1800	8200	3900	15000	3.5
2.4	2.1	1800	8200	1800	18000	1.5
2.4	2.2	1800	8200	1500	18000	1.3
2.4	2.3	1800	8200	1200	18000	1.1
2.6	0.9	2200	8200	18000	100	16.7
2.6	1.0	2200	8200	15000	2700	14.2
2.6	1.2	2200	8200	12000	5600	11.4

SCHMITT TRIGGER CIRCUITS

GERMANIUM 9 VOLT

VR	VT	RE	R2	RL	R1	LG
2.6	1.3	2200	8200	10000	8200	9.2
2.6	1.4	2200	8200	8200	10000	7.6
2.6	1.7	2200	8200	5600	12000	5.3
2.6	2.1	2200	8200	2700	15000	2.6
2.8	1.1	2200	8200	15000	1000	15.5
2.8	1.2	2200	8200	12000	3900	12.5
2.8	1.3	2200	8200	10000	5600	10.6
2.8	1.8	2200	8200	5600	10000	5.9
2.8	2.0	2200	8200	3900	12000	4.0
2.8	2.6	2200	8200	1200	15000	1.2
2.8	2.7	2200	8200	1000	15000	1.0
3.0	1.4	2700	8200	12000	2700	13.4
3.0	1.5	2700	8200	10000	4700	11.2
3.0	1.7	2700	8200	8200	6800	9.0
3.0	1.8	2700	8200	6800	8200	7.5
3.0	2.1	2700	8200	4700	10000	5.2
3.0	2.5	2700	8200	2700	12000	3.0
3.2	1.4	2700	8200	12000	1200	14.7
3.2	1.6	2700	8200	10000	3300	12.1
3.2	1.9	2700	8200	6800	6800	8.1
3.2	2.4	2700	8200	3300	10000	4.0
3.2	2.9	2700	8200	1500	12000	1.8
3.2	3.0	2700	8200	1200	12000	1.5
3.4	1.4	2700	10000	12000	2700	14.1
3.4	1.6	2700	10000	10000	4700	11.8
3.4	2.3	2700	10000	4700	10000	5.5
3.4	2.7	2700	10000	2700	12000	3.2
3.4	2.9	2700	10000	2200	12000	2.7
3.6	1.6	3300	10000	12000	1500	15.2
3.6	1.8	3300	10000	10000	3300	12.8
3.6	2.2	3300	10000	6800	6800	8.6
3.6	2.8	3300	10000	3300	10000	4.2
3.6	3.3	3300	10000	1500	12000	1.9
3.6	3.4	3300	10000	1200	12000	1.5
3.8	1.7	3300	10000	12000	270	16.4
3.8	1.9	3300	10000	10000	2200	13.8
3.8	2.1	3300	10000	8200	3900	11.4
3.8	2.2	3300	10000	6800	5600	9.2
3.8	2.4	3300	10000	5600	6800	7.6
3.8	2.8	3300	10000	3900	8200	5.4

VR	VT	RE	R2	RL	R1	LG
3.8	3.2	3300	10000	2200	10000	3.0
4.0	1.9	3300	10000	10000	1200	14.7
4.0	2.5	3300	10000	5600	5600	8.3
4.0	3.7	3300	10000	1200	10000	1.8
4.0	3.8	3300	10000	1000	10000	1.5
4.2	1.9	3900	12000	12000	270	17.2
4.2	2.1	3900	12000	10000	2200	14.4
4.2	2.3	3900	12000	8200	3900	11.9
4.2	2.5	3900	12000	6800	5600	9.6
4.2	2.7	3900	12000	5600	6800	7.9
4.2	3.1	3900	12000	3900	8200	5.6
4.2	3.6	3900	12000	2200	10000	3.2
4.4	2.1	3900	12000	10000	1200	15.4
4.4	2.8	3900	12000	5600	5600	8.6
4.4	4.1	3900	12000	1200	10000	1.8
4.4	4.2	3900	12000	1000	10000	1.6
4.6	2.2	3900	12000	10000	270	16.4
4.6	2.4	3900	12000	8200	2200	13.4
4.6	2.6	3900	12000	6800	3300	11.3
4.6	2.9	3900	12000	5600	4700	9.2
4.6	3.1	3900	12000	4700	5600	7.7
4.6	3.5	3900	12000	3300	6800	5.5
4.6	3.8	3900	12000	2200	8200	3.6
4.8	2.5	3900	12000	8200	1200	14.4
4.8	2.7	3900	12000	6800	2700	11.9
4.8	2.9	3900	12000	5600	3900	9.8
4.8	3.1	3900	12000	4700	4700	8.3
4.8	3.3	3900	12000	3900	5600	6.8
4.8	3.7	3900	12000	2700	6800	4.7
4.8	4.4	3900	12000	1200	8200	2.1
4.8	4.5	3900	12000	1000	8200	1.8
5.0	2.7	4700	12000	8200	560	15.3
5.0	3.0	4700	12000	6800	1800	12.8
5.0	3.2	4700	12000	5600	3300	10.3
5.0	3.4	4700	12000	4700	3900	8.9
5.0	3.6	4700	12000	3900	4700	7.4
5.0	3.8	4700	12000	3300	5600	6.1
5.0	4.4	4700	12000	1800	6800	3.4
5.0	4.8	4700	12000	680	8200	1.3
5.0	4.9	4700	12000	560	8200	1.0

SCHMITT TRIGGER CIRCUITS

GERMANIUM 9 VOLT

VR	VT	RE	R2	RL	R1	LG
5.2	2.8	4700	15000	8200	1500	14.7
5.2	4.5	4700	15000	1800	8200	3.2
5.2	4.7	4700	15000	1500	8200	2.7
5.4	2.8	4700	15000	8200	680	15.7
5.4	3.1	4700	15000	6800	2200	12.9
5.4	3.4	4700	15000	5600	3300	10.7
5.4	4.0	4700	15000	3300	5600	6.3
5.4	4.4	4700	15000	2200	6800	4.2
5.4	5.1	4700	15000	820	8200	1.6
5.4	5.2	4700	15000	680	8200	1.3
5.6	2.9	4700	15000	8200	100	16.5
5.6	3.1	4700	15000	6800	1500	13.7
5.6	3.2	4700	15000	6800	1200	14.0
5.6	3.4	4700	15000	5600	2700	11.3
5.6	3.7	4700	15000	4700	3300	9.7
5.6	4.1	4700	15000	3300	4700	6.8
5.6	4.3	4700	15000	2700	5600	5.4
5.6	4.9	4700	15000	1500	6800	3.0
5.6	5.1	4700	15000	1200	6800	2.5
5.8	2.9	4700	18000	8200	470	16.4
5.8	3.2	4700	18000	6800	1800	13.7
5.8	3.5	4700	18000	5600	3300	11.0
5.8	3.8	4700	18000	4700	3900	9.5
5.8	4.0	4700	18000	3900	4700	7.8
5.8	4.2	4700	18000	3300	5600	6.5
5.8	4.9	4700	18000	1800	6800	3.6
5.8	5.7	4700	18000	560	8200	1.1
6.0	3.5	5600	18000	6800	1200	14.4
6.0	4.1	5600	18000	4700	3300	10.0
6.0	4.5	5600	18000	3300	4700	7.0
6.0	5.5	5600	18000	1200	6800	2.5
6.2	3.6	5600	18000	6800	390	15.5
6.2	3.9	5600	18000	5600	1500	12.9
6.2	4.1	5600	18000	4700	2700	10.5
6.2	4.4	5600	18000	3900	3300	8.9
6.2	4.6	5600	18000	3300	3900	7.5
6.2	4.9	5600	18000	2700	4700	6.1
6.2	5.3	5600	18000	1800	5600	4.0
6.2	5.5	5600	18000	1500	5600	3.5
6.2	6.0	5600	18000	560	6800	1.3

SCHMITT TRIGGER CIRCUITS

GERMANIUM 12 VOLT

VR	VT	RE	R2	RL	R1	LG
1.00	0.75	680	6800	6800	47000	2.0
1.25	0.75	1000	6800	12000	33000	4.4
1.25	1.00	1000	6800	5600	39000	2.1
1.50	0.75	1200	6800	18000	22000	7.5
1.50	1.00	1200	6800	6800	33000	2.8
1.50	1.25	1200	6800	4700	33000	2.1
1.75	0.75	1500	6800	22000	12000	10.7
1.75	1.00	1500	6800	12000	22000	5.9
1.75	1.25	1500	6800	6800	27000	3.3
1.75	1.50	1500	6800	3300	33000	1.5
2.00	0.75	1500	6800	22000	6800	12.5
2.00	1.00	1500	6800	12000	18000	6.6
2.00	1.25	1500	6800	8200	22000	4.5
2.00	1.50	1500	6800	5600	22000	3.3
2.00	1.75	1500	6800	2700	27000	1.5
2.25	0.75	1800	6800	27000	100	16.4
2.25	1.00	1800	6800	18000	8200	11.3
2.25	1.25	1800	6800	10000	15000	6.5
2.25	1.50	1800	6800	6800	18000	4.5
2.25	1.75	1800	6800	3900	22000	2.5
2.25	2.00	1800	6800	3300	22000	2.1
2.50	1.00	2200	6800	22000	1200	15.5
2.50	1.25	2200	6800	15000	8200	10.5
2.50	1.50	2200	6800	8200	15000	5.8
2.50	1.75	2200	6800	5600	18000	3.9
2.50	2.00	2200	6800	4700	18000	3.4
2.50	2.25	2200	6800	2200	22000	1.5
2.75	1.00	2200	8200	22000	2200	15.8
2.75	1.25	2200	8200	15000	10000	10.5
2.75	1.50	2200	8200	10000	15000	7.0
2.75	1.75	2200	8200	6800	18000	4.8
2.75	2.00	2200	8200	5600	18000	4.1
2.75	2.25	2200	8200	3300	22000	2.3
2.75	2.50	2200	8200	2200	22000	1.6
3.00	1.25	2700	8200	18000	3900	14.2
3.00	1.50	2700	8200	15000	6800	11.9
3.00	1.75	2700	8200	10000	12000	7.9
3.00	2.00	2700	8200	6800	15000	5.4
3.00	2.25	2700	8200	4700	18000	3.6
3.00	2.50	2700	8200	3900	18000	3.1

SCHMITT TRIGGER CIRCUITS

GERMANIUM 12 VOLT

VR	VT	RE	R2	RL	R1	LG
3.25	1.25	2700	8200	18000	1800	15.5
3.25	1.50	2700	8200	15000	4700	13.0
3.25	1.75	2700	8200	10000	10000	8.6
3.25	2.00	2700	8200	8200	12000	7.0
3.25	2.25	2700	8200	5600	15000	4.7
3.25	2.50	2700	8200	4700	15000	4.1
3.25	2.75	2700	8200	2700	18000	2.2
3.25	3.00	2700	8200	1800	18000	1.6
3.50	1.25	2700	8200	18000	100	16.8
3.50	1.50	2700	8200	15000	2700	14.3
3.50	1.75	2700	8200	12000	5600	11.5
3.50	2.00	2700	8200	8200	10000	7.6
3.50	2.25	2700	8200	6800	12000	6.1
3.50	2.50	2700	8200	5600	12000	5.4
3.50	2.75	2700	8200	3300	15000	3.1
3.50	3.00	2700	8200	2700	15000	2.6
3.75	1.75	3300	8200	15000	1500	15.2
3.75	2.00	3300	8200	12000	4700	12.1
3.75	2.25	3300	8200	8200	8200	8.4
3.75	2.50	3300	8200	6800	10000	6.8
3.75	2.75	3300	8200	4700	12000	4.7
3.75	3.00	3300	8200	3900	12000	4.1
3.75	3.25	3300	8200	2200	15000	2.2
3.75	3.50	3300	8200	1500	15000	1.5
4.00	1.75	3300	8200	15000	100	16.4
4.00	2.00	3300	8200	12000	2700	13.5
4.00	2.25	3300	8200	10000	4700	11.2
4.00	2.50	3300	8200	6800	8200	7.5
4.00	2.75	3300	8200	5600	10000	6.0
4.00	3.00	3300	8200	4700	10000	5.3
4.00	3.25	3300	8200	3300	12000	3.6
4.00	3.50	3300	8200	2700	12000	3.0
4.25	2.25	3900	8200	12000	1800	14.2
4.25	2.50	3900	8200	10000	3900	11.8
4.25	2.75	3900	8200	8200	5600	9.7
4.25	3.00	3900	8200	5600	8200	6.6
4.25	3.25	3900	8200	3900	10000	4.6
4.25	3.50	3900	8200	3300	10000	4.0
4.25	3.75	3900	8200	2200	12000	2.5
4.25	4.00	3900	8200	1500	12000	1.8

SCHMITT TRIGGER CIRCUITS
GERMANIUM 12 VOLT

VR	VT	RE	R2	RL	R1	LG
4.50	2.00	3900	10000	15000	220	17.3
4.50	2.25	3900	10000	12000	3300	13.7
4.50	2.50	3900	10000	10000	5600	11.3
4.50	2.75	3900	10000	8200	6800	9.5
4.50	3.00	3900	10000	6800	8200	7.9
4.50	3.75	3900	10000	3300	12000	3.8
4.50	4.00	3900	10000	2700	12000	3.2
4.75	2.25	3900	10000	12000	1800	15.0
4.75	2.50	3900	10000	10000	3900	12.4
4.75	2.75	3900	10000	8200	5600	10.2
4.75	3.00	3900	10000	6800	6800	8.6
4.75	3.25	3900	10000	5600	8200	7.0
4.75	3.75	3900	10000	3900	10000	4.8
4.75	4.00	3900	10000	2200	12000	2.7
4.75	4.25	3900	10000	1800	12000	2.2
4.75	4.50	3900	10000	1500	12000	1.9
5.00	2.50	4700	10000	12000	1000	15.8
5.00	3.00	4700	10000	8200	4700	10.8
5.00	3.75	4700	10000	4700	8200	6.2
5.00	4.00	4700	10000	3300	10000	4.3
5.00	4.25	4700	10000	2700	10000	3.6
5.00	4.50	4700	10000	1500	12000	1.9
5.00	4.75	4700	10000	1200	12000	1.6
5.25	2.75	4700	10000	12000	100	16.7
5.25	3.00	4700	10000	10000	1800	14.2
5.25	3.25	4700	10000	8200	3300	11.9
5.25	3.50	4700	10000	6800	4700	9.9
5.25	4.00	4700	10000	3900	8200	5.4
5.25	4.25	4700	10000	3300	8200	4.8
5.25	4.50	4700	10000	2200	10000	3.0
5.25	4.75	4700	10000	1800	10000	2.6
5.25	5.00	4700	10000	1500	10000	2.2
5.50	3.00	4700	10000	10000	820	15.2
5.50	3.25	4700	10000	8200	2700	12.4
5.50	3.50	4700	10000	6800	3900	10.5
5.50	3.75	4700	10000	5600	5600	8.3
5.50	4.25	4700	10000	3900	6800	6.0
5.50	4.50	4700	10000	2700	8200	4.1
5.50	5.00	4700	10000	1200	10000	1.8
5.50	5.25	4700	10000	1000	10000	1.5

SCHMITT TRIGGER CIRCUITS

GERMANIUM 12 VOLT

VR	VT	RE	R2	RL	R1	LG
5.75	2.75	4700	12000	12000	100	17.4
5.75	3.00	4700	12000	10000	1800	14.8
5.75	3.25	4700	12000	8200	3900	11.9
5.75	3.50	4700	12000	6800	5600	9.7
5.75	3.75	4700	12000	6800	4700	10.3
5.75	4.25	4700	12000	3900	8200	5.7
5.75	5.00	4700	12000	2200	10000	3.2
5.75	5.25	4700	12000	1800	10000	2.7
6.00	3.25	5600	12000	10000	1200	15.5
6.00	3.50	5600	12000	10000	1000	15.7
6.00	3.75	5600	12000	8200	2700	12.9
6.00	4.00	5600	12000	6800	3900	10.9
6.00	4.25	5600	12000	5600	5600	8.7
6.00	4.50	5600	12000	4700	6800	7.1
6.00	4.75	5600	12000	3900	6800	6.2
6.00	5.00	5600	12000	2700	8200	4.3
6.00	5.50	5600	12000	1500	10000	2.3
6.00	5.75	5600	12000	1000	10000	1.6
6.25	3.50	5600	12000	10000	180	16.6
6.25	3.75	5600	12000	8200	2200	13.4
6.25	4.00	5600	12000	6800	3300	11.4
6.25	4.25	5600	12000	5600	4700	9.2
6.25	4.50	5600	12000	4700	5600	7.8
6.25	5.00	5600	12000	3300	6800	5.5
6.25	5.50	5600	12000	2200	8200	3.6
6.25	5.75	5600	12000	1800	8200	3.0
6.50	3.75	5600	12000	8200	1200	14.5
6.50	4.25	5600	12000	6800	2700	11.9
6.50	4.50	5600	12000	5600	3900	9.8
6.50	4.75	5600	12000	4700	4700	8.3
6.50	5.00	5600	12000	3900	5600	6.8
6.50	5.50	5600	12000	2700	6800	4.7
6.50	5.75	5600	12000	2200	6800	4.0
6.50	6.00	5600	12000	1200	8200	2.1
6.50	6.25	5600	12000	1000	8200	1.8
6.75	4.00	5600	12000	8200	390	15.5
6.75	4.25	5600	12000	6800	1800	12.9
6.75	4.50	5600	12000	5600	3300	10.3
6.75	4.75	5600	12000	4700	3900	8.9
6.75	5.00	5600	12000	3900	4700	7.4

SCHMITT TRIGGER CIRCUITS
GERMANIUM 24 VOLT

VR	VT	RE	R2	RL	R1	LG
1.5	1.0	1200	5600	15000	56000	3.3
2.0	1.0	1500	6800	27000	39000	7.0
2.0	1.5	1500	6800	10000	56000	2.6
2.5	1.0	2200	6800	47000	5600	15.4
2.5	1.5	2200	6800	18000	33000	6.1
2.5	2.0	2200	6800	6800	47000	2.2
3.0	1.0	2700	6800	39000	4700	15.3
3.0	1.5	2700	6800	27000	18000	10.3
3.0	2.0	2700	6800	15000	27000	6.1
3.0	2.5	2700	6800	5600	39000	2.2
3.5	1.5	2700	6800	33000	3300	15.5
3.5	2.0	2700	6800	18000	18000	8.5
3.5	2.5	2700	6800	10000	27000	4.6
3.5	3.0	2700	6800	4700	33000	2.1
4.0	1.5	3300	6800	33000	100	16.9
4.0	2.0	3300	6800	22000	10000	11.6
4.0	2.5	3300	6800	15000	15000	8.4
4.0	3.0	3300	6800	10000	22000	5.3
4.0	3.5	3300	6800	4700	27000	2.5
4.5	2.0	3900	6800	27000	680	16.4
4.5	2.5	3900	6800	22000	5600	13.4
4.5	3.0	3900	6800	12000	15000	7.4
4.5	3.5	3900	6800	6800	22000	4.0
4.5	4.0	3900	6800	4700	22000	2.9
5.0	2.5	4700	6800	22000	2700	14.8
5.0	3.0	4700	6800	18000	6800	12.1
5.0	3.5	4700	6800	12000	12000	8.3
5.0	4.0	4700	6800	6800	18000	4.6
5.0	4.5	4700	6800	3300	22000	2.2
5.5	3.0	4700	6800	18000	3900	13.5
5.5	3.5	4700	6800	15000	6800	11.3
5.5	4.0	4700	6800	10000	12000	7.5
5.5	4.5	4700	6800	6800	15000	5.1
5.5	5.0	4700	6800	3300	18000	2.5
6.0	3.0	5600	8200	22000	1200	16.7
6.0	3.5	5600	8200	18000	5600	13.5
6.0	4.0	5600	8200	15000	8200	11.4
6.0	4.5	5600	8200	8200	15000	6.2
6.0	5.0	5600	8200	5600	18000	4.2
6.0	5.5	5600	8200	2200	22000	1.6

SCHMITT TRIGGER CIRCUITS

GERMANIUM 24 VOLT

VR	VT	RE	R2	RL	R1	LG
6.5	3.5	5600	8200	18000	3300	14.8
6.5	4.0	5600	8200	15000	5600	12.7
6.5	4.5	5600	8200	10000	12000	8.0
6.5	5.0	5600	8200	8200	12000	7.0
6.5	5.5	5600	8200	5600	15000	4.7
6.5	6.0	5600	8200	2700	18000	2.3
7.0	4.0	5600	8200	15000	3900	13.7
7.0	4.5	5600	8200	12000	6800	11.0
7.0	5.0	5600	8200	10000	8200	9.4
7.0	5.5	5600	8200	6800	12000	6.2
7.0	6.0	5600	8200	3900	15000	3.6
7.0	6.5	5600	8200	1800	18000	1.6
7.5	4.0	6800	8200	18000	100	17.1
7.5	4.5	6800	8200	15000	2200	14.9
7.5	5.0	6800	8200	12000	5600	11.7
7.5	5.5	6800	8200	10000	6800	10.1
7.5	6.0	6800	8200	6800	10000	6.9
7.5	6.5	6800	8200	4700	12000	4.8
7.5	7.0	6800	8200	2200	15000	2.2
8.0	5.0	6800	8200	15000	680	16.1
8.0	5.5	6800	8200	10000	5600	10.8
8.0	6.0	6800	8200	8200	6800	9.1
8.0	6.5	6800	8200	5600	10000	6.0
8.0	7.0	6800	8200	3900	12000	4.1
8.5	5.5	6800	8200	12000	2200	14.0
8.5	6.0	6800	8200	10000	3900	11.9
8.5	7.0	6800	8200	4700	10000	5.3
8.5	7.5	6800	8200	3900	10000	4.6
8.5	8.0	6800	8200	2200	12000	2.6
9.0	5.5	8200	10000	15000	1000	16.7
9.0	6.0	8200	10000	12000	3900	13.4
9.0	6.5	8200	10000	10000	5600	11.4
9.0	7.5	8200	10000	5600	10000	6.4
9.0	8.0	8200	10000	3900	12000	4.4
9.0	8.5	8200	10000	1500	15000	1.6
9.5	5.5	8200	10000	15000	100	17.5
9.5	6.5	8200	10000	10000	4700	12.0
9.5	7.0	8200	10000	8200	6800	9.6
9.5	7.5	8200	10000	6800	8200	8.0
9.5	8.0	8200	10000	4700	10000	5.6

SCHMITT TRIGGER CIRCUITS

GERMANIUM 24 VOLT

VR	VT	RE	R_2	RL	R_1	LG
9.5	8.5	8200	10000	2700	12000	3.2
9.5	9.0	8200	10000	2200	12000	2.7
10.0	6.5	8200	10000	12000	1200	15.7
10.0	7.0	8200	10000	10000	3300	13.0
10.0	7.5	8200	10000	6800	6800	8.7
10.0	8.0	8200	10000	5600	8200	7.1
10.0	8.5	8200	10000	4700	8200	6.3
10.0	9.0	8200	10000	3300	10000	4.3
10.0	9.5	8200	10000	1500	12000	1.9
10.5	7.0	10000	10000	12000	390	16.5
10.5	7.5	10000	10000	10000	2700	13.5
10.5	8.0	10000	10000	8200	3900	11.5
10.5	8.5	10000	10000	6800	5600	9.4
10.5	9.5	10000	10000	2700	10000	3.6
10.5	10.0	10000	10000	2200	10000	3.1
11.0	8.0	10000	10000	10000	1200	14.9
11.0	8.5	10000	10000	6800	4700	9.9
11.0	9.0	10000	10000	5600	5600	8.4
11.0	9.5	10000	10000	4700	6800	6.9
11.0	10.0	10000	10000	3300	8200	4.8
11.0	10.5	10000	10000	1500	10000	2.2
11.5	7.5	10000	12000	12000	470	17.1
11.5	8.0	10000	12000	10000	2700	14.1
11.5	8.5	10000	12000	8200	4700	11.4
11.5	9.0	10000	12000	6800	5600	9.7
11.5	9.5	10000	12000	5600	6800	8.0
11.5	10.0	10000	12000	3900	8200	5.7
11.5	10.5	10000	12000	2700	10000	3.8
11.5	11.0	10000	12000	1000	12000	1.4
12.0	8.5	10000	12000	10000	1500	15.2
12.0	9.0	10000	12000	8200	3300	12.5
12.0	10.0	10000	12000	4700	6800	7.2
12.0	10.5	10000	12000	3300	8200	5.0
12.0	11.0	10000	12000	1800	10000	2.7
12.0	11.5	10000	12000	1500	10000	2.3
12.5	9.0	12000	12000	10000	560	16.3
12.5	9.5	12000	12000	8200	2200	13.5
12.5	10.0	12000	12000	6800	3900	11.0
12.5	10.5	12000	12000	5600	4700	9.3
12.5	11.0	12000	12000	3900	6800	6.3

SCHMITT TRIGGER CIRCUITS
GERMANIUM 24 VOLT

VR	VT	RE	R2	RL	R1	LG
12.5	11.5	12000	12000	2700	8200	4.3
12.5	12.0	12000	12000	1000	10000	1.6
13.0	9.0	12000	12000	10000	100	16.9
13.0	9.5	12000	12000	8200	1800	13.9
13.0	10.0	12000	12000	8200	1500	14.3
13.0	10.5	12000	12000	6800	2700	12.0
13.0	11.0	12000	12000	4700	4700	8.4
13.0	11.5	12000	12000	3900	5600	6.9
13.0	12.0	12000	12000	2700	6800	4.8
13.0	12.5	12000	12000	1500	8200	2.6
13.5	10.0	12000	12000	8200	820	15.1
13.5	10.5	12000	12000	6800	2200	12.5
13.5	11.0	12000	12000	5600	3300	10.4
13.5	12.0	12000	12000	3300	5600	6.1
13.5	12.5	12000	12000	2200	6800	4.0
13.5	13.0	12000	12000	1000	8200	1.8
14.0	9.5	12000	15000	10000	220	17.4
14.0	10.5	12000	15000	8200	1800	14.5
14.0	11.0	12000	15000	6800	3300	11.9
14.0	11.5	12000	15000	4700	5600	8.1
14.0	12.5	12000	15000	3300	6800	5.8
14.0	13.0	12000	15000	2200	8200	3.8
14.0	13.5	12000	15000	1800	8200	3.2
14.5	10.5	12000	15000	8200	1200	15.2
14.5	11.0	12000	15000	6800	2700	12.5
14.5	11.5	12000	15000	5600	3900	10.3
14.5	12.0	12000	15000	4700	4700	8.7
14.5	12.5	12000	15000	3900	5600	7.2
14.5	13.0	12000	15000	2700	6800	5.0
14.5	13.5	12000	15000	2200	6800	4.2
14.5	14.0	12000	15000	1200	8200	2.2
15.0	11.0	12000	15000	8200	390	16.2
15.0	11.5	12000	15000	6800	1800	13.4
15.0	12.0	12000	15000	5600	2700	11.4
15.0	12.5	12000	15000	4700	3900	9.3
15.0	13.0	12000	15000	3900	4700	7.7
15.0	13.5	12000	15000	2700	5600	5.5
15.0	14.0	12000	15000	1800	6800	3.6
15.0	14.5	12000	15000	680	8200	1.3
15.5	12.0	15000	15000	6800	1200	14.2